CHALLENGEABILITY IN MODERN SCIENCE

Challengeability in Modern Science

J. O. WISDOM

Avebury

Aldershot · Brookfield USA · Hong Kong · Singapore · Sydney

© J.O. Wisdom, 1987

All rights reserved. No part of this publication may be reproduced, stored in a retrieval system, or transmitted in any form or by any means, electronic, mechanical, photocopying, recording, or otherwise without the prior permission of Gower Publishing Company Limited.

Published by
Gower Publishing Company Limited
Gower House
Croft Road
Aldershot
Hants GU11 3HR
England

Gower Publishing Company
Old Post Road
Brookfield
Vermont 05036
USA

Printed and bound in Great Britain by
Blackmore Press, Shaftesbury, Dorset

British Library Cataloguing in Publication Data

Wisdom, J.O.
   Challengeability in modern science.
   I. Title
   501   Q175

ISBN 0 566 05026 9

Contents

	Preface	vii
	Reprinting Acknowledgements	x

Part I: Philosophical problems underlying modern science

1	Introduction: The role of philosophy of science	3
2	The origin of modern science: observationalism and induction contra essentialism	8
	Appendix: Known objections to induction	12
3	Four contemporary interpretations of the nature of science	16
	Appendix: Meaninglessness in quantum mechanics	27
4	The role of 'truth' in science	32
5	The common problem: the observation–theory gap	36
6	The demarcation problem	41
	Appendix: Popper's metascience of explanation	45
7	Theories not absolutely falsifiable	49
8	The nature of 'normal' science	62

Part II: Observationalism

9	Observations *not* the building blocks of science: the structure of science as theories of layered observations	71
	Appendix: Science as a form of currency reserve	74
10	The metaphysics of observationalism: the heir to essentialism	76
11	After observationalism — whither?	81
12	Observations as theory-laden	83
	Appendix 1: The fallibility of observation — a medical example	87
	Appendix 2: *Die Philosophiedämmerung*	89
13	The conception of 'theory' as 'non-instantiative'	92

Part III: Reality, weltanschauungen and discovery

14	Conventionalism, truth, and cosmological furniture	107
	Appendix: The incommensurability thesis	120
15	Scientific theory: empirical content, embedded ontology, and weltanschauung	125
	Appendix 1: The demarcation problem	138
	Appendix 2: The notion of a model	140
16	Weltanschauungen as sources of our knowledge and our ignorance	146
17	The handmaid — science or metaphysics?	156
	Index	163

Preface

This book provides the essentials from philosophy of natural science, which are generally needed for close study of philosophy of the social sciences.

One major aim of this book is to articulate the role in science of philosophy of science, which is not normally discussed. There are times when science itself can get bogged down by an erroneous approach to it, that is, by a faulty philosophy of science. This is a century in which there have been great advances in theoretical science, at least in the first third of it or so; at the same time, at the hands of famous scientists themselves, philosophies of science have sprung up which may go some way to explaining the falling off in theoretical advance. In particular the fear of deviating from what was conceived as being the true-blue nature and procedure of science has led to an attitude, it is claimed here, of being perpetually on guard against backsliding into the anti-scientific ideas of the middle ages. Further, scientists have adopted interpretations of the nature of science which prevent some important questions from being raised. And in desperation they mostly even regard the question of the truth of theories as of minor importance.

Part I is concerned with sorting out these views of science and to show that the concept of truth is central with all that this carries in its train.

Part II is devoted to the 400-year-old Philosophy of Observationalism. Thence to the idea that observations are theory-laden. And to the peculiar nature of scientific theories.

Part III is concerned with the unfashionable question of what sorts of things there may be in the universe and whether they can be reasonably claimed to exist. For the question of truth cannot be answered along realistic lines if we have no way of knowing what exists.

It transpires further that in science we deal not only with the content of theories, which tell us how things work in the world — regarded usually as the sole function of science — but with the basic structure of the world, here called embedded ontology. The significant point that emerges here is that this is a form of metaphysics which is not susceptible of any kind of scientific proof.

Along with this there transpires another ingredient — the weltanschauung of a scientific theory — seen not as an excrescence which competent successors will eliminate, but as part of the life-blood of science itself, determining what shall and what shall not count as science, fostering its growth or just as often impeding it. Here is another form of metaphysics, long dimly recognised though not sharply detached for scrutiny, underlying science. This additional form leads to a significant new question about the role of philosophy — or more specifically of weltanschauungen — in scientific progress, and about the unique difficulty of tackling this factor. It is pointed out that the difficulty hinges on its connection with creativity with all the mysteriousness of this phenomenon; for to become self-conscious about the phenomenon is to take the first step towards coming to grips with it.

The book is not intended for beginners, but it is written reasonably simply and could be understood by a beginner who had some help. Although there is much that is new in it, it should be within the compass of a senior undergraduate student.

As always I owe much to Sir Karl Popper's teaching. What is put forward here, though not incompatible with Sir Karl's philosophy of science, diverges slightly from it, and is not foreshadowed by it (though he has made some undeveloped allusions pointing in the same general direction).

If I may acknowledge Sir Karl's great innovation in this field, perhaps the best tribute I can make to it is to try to go beyond it by building on it.

Parts of the book have been written over many years. My chief debt is to my students, mainly at York University, Toronto, but also at the University of Southern California and at the State University of New York at Fredonia. In all these, the quality of questions or criticisms was exhilarating and profitable to me, and I hope to the students. I would mention in particular Dr. T.C. Skate of USC, and Dr Catherine May and Dr Tom Bowers of York University, Toronto; and earlier at the London School of Economics and Political Science, Mrs Helena Sheiham; without such students I would have been pottering around at routine classroom lore. Dr May wrote me one of the best papers I have ever had from an undergraduate, a superb criticism of my paper, now the substance of Chapter 15 on ontology and weltanschauung. I am also indebted, as for most things, to my friends and colleagues at York, Professor Ian Jarvie and Professor Jagdish Hattiangadi, whose help and advice was always readily available. But most of all to a surprising source. My publisher's reader made a number of suggestions both general and detailed. Not only did they show understanding, but also practicality and a flair for improvement; I was more than glad to fall in with almost every one of his suggestions. Among them were proposals for cuts and recombination. I carried out a few of these myself. But most of this work was done by my wife, Clara, who had been teaching philosophy of science. The remarkable skill she brought to bear consisted in cutting, joining up the segments, but doing so in the style of the original, in such a way that no one, not even I myself, would notice that a cut and junction had been made; she also made other valuable changes.

I wish to make special mention of a work by Professor Håkon Törnebohm's, *Paradigm shift in the theories of gravitation*, Norstelts Tryckeri AB, Stockholm, 1977. Professor Törnebohm provides the mathematics of, or as much of it as is relevant to, his problems, beautifully done, for example, on the relation of general relativity to Newtonian and Keplerian derivations (the title is a bit misleading as it wrongly suggests he is dealing with Kuhn). His ideas are remarkably similar to mine, and I can only regret that I did not know about his work before I had written the corresponding parts of mine. His work and mine might well make a suitable pair of companion volumes.

I go against the usual printing convention of italicising foreign words. Because I wish to retain italics exclusively for emphasis, I have requested that all foreign words should be printed in roman type. Also I write German nouns without capitals.

Wilmont House
Castlebridge
Co. Wexford

J.O.Wisdom

23 April 1986

Acknowledgements

The following papers are reprinted by kind permission in whole or in part, with minor corrections:

'Four Contemporary Interpretations of the Nature of Science', *Foundations of Physics*, ed. Yourgrau, Plenum, New York, 1, pp.269–84, 1971, reproduced in Chapter 3 and in the Appendix to Chapter 2.

'Observations as the Building Blocks of Science in 20th Century Scientific Thought', *Boston Studies in the Philosophy of Science*, ed. Cohen and Wartoffsky, Reidel, Dordrecht, Vol.8, pp.212–22, 1972, reproduced in Chapter 9 and its Appendix.

'Scientific Theory: Empirical Content, Embedded Ontology, and Weltanschauung', *Philosophy and Phenomenological Research*, 33, pp.62–77, 1972, reproduced by permission of the editor in Chapter 15.

(With Alan Thompson) 'The Fallibility of Observation', *Dialogue*, 13, pp.353–4, 1974, reproduced by permission of the editor of *Dialogue*, also of the late Professor Alan Thompson and of his son, Mr Piers Thompson, as Appendix 1 to Chapter 12.

'The Incommensurability Thesis', *Philosophical Studies*, 25, pp.299–301, 1974, reproduced by permission of the editor of Reidel, Dordrecht, as an Appendix to Chapter 14.

'The Nature of "Normal" Science', *The Philosophy of Karl Popper*, ed. Schilpp, Open Court, Chicago, Vol.2, pp.820–42, 1974, reproduced in Chapters 7 and 8.

'Conventionalism, Truth, and Cosmological Furniture', *Canadian Journal of Philosophy*, 4, pp.441–57, 1975, reproduced by permission of the Canadian Association for Publishing in Philosophy, in Chapter 14.

PART I: PHILOSOPHICAL PROBLEMS UNDERLYING MODERN SCIENCE

1 Introduction: The role of philosophy of science

One major aim in this book is to articulate the role of the philosophy of science in doing science itself, by highlighting the metaphysical components of scientific theories, and raising new or disregarded questions by challenging frameworks. A cognate aim is to show that philosophy of science also functions at times so as to censor important questions and damage the progress of science, not by distorting results, but by restricting the area of scientific problems studied.

The second major aim is to portray modern science as having a root in the anti-essentialism which led to its origin 400 years ago, and also a root in the belief that observations are the bedrock of science. Most of the outstanding scientists of this century have seen science in this light. Not only is it contended that modern science developed from the Renaissance as a consequence of breaking away from essentiality (see Popper, 1946, 1963); it is further claimed here that the very fabric of science is to be understood as anti-essentialistic.[1]

This is a century in which there were great advances in theoretical science, at least in the first third of it or so; at the same time, at the hands of scientists themselves, philosophies of science sprang up which partly explain the falling off in theoretical advance. In particular, the fear of deviating from what was conceived as being the true-blue nature and procedure of science has led to an attitude of being perpetually on guard against backsliding into the anti-scientific ideas of the middle ages. Thus interpretations of science, as found in some outstanding scientists of this century, reveal both the anti-essentialistic leitmotif and also the acceptance of observationalism as a sure, and indeed the only sure, foundation for the development of scientific theory.

In desperation some of these scientists/philosophers of science have even regarded the question of the truth of theories as of minor importance.

New arguments are brought forward so as vastly to reduce the area of the conflicting positions, that is, to resolve the issue of varying approaches to science, and to show that the concept of truth is central in interpreting science, with all that this carries in its train.

It thus becomes necessary to free the interpretation of science from an over-reaction to essentialism — even to the extent of denying the role of the concept of truth. Here the springboard is Popper's radical re-interpretation based on the falsifiability principle. Because of widespread past and present misinterpretation of Popper, reflected by Kuhn, it is necessary to set the record straight, to the effect that Popper never held refutability to be absolute. However, this opens up a problem. Two consequences of Popper's view necessitate widespread reorientations. One is that falsifiability is in principle a relative falsifiability, so that while we cannot confirm scientific theories at all, we are restricted in our power to falsify them. The other consequence is that the absoluteness of observation has to be given up; and here new argument is adduced for this conclusion. Observations are to be regarded as theory-laden, as interpretations. But with theories somewhat shaky as regards truth-value, and now also observations, then how do we determine a theory's truth-value?

With the doubts attending the truth-value of theories and observations, conventionalism is condoned with new vigour — as was noted above, the controversial issue concerning interpretations of science could not be totally dispelled, for despite strong undermining, conventionalism remained just possibly viable. Thus the third aim is to come to grips finally with conventionalism. The attempt to cope with this problem leads to a new way of trying to handle the correspondence theory of truth, to satisfy two desiderata: one is to avoid the unsupported 'realism' postulated though not discussed by Popper; the other is to provide an approximation to it, taking cognisance of the non-absoluteness of observation.

It will be seen that what is in question is the status of science as a rational enterprise. The risks involved are twofold. One is that if doubt is cast on the truth-value of theories and also on the truth-value of observations, we are brought to the brink of scepticism (which, of course, is the outcome of two decades of philosophy of science by Kuhn, Feyerabend, and Lakatos), with the threat of reverting to the essentialism of the middle ages. It seems to me, however, that by making certain additions to Popper, both essentialism and scepticism can be met within the framework of objectivity.[2] That is to say,

Popper's refutability approach can be made to work, first by developing the maximally powerful arguments against certain alternatives (notably against induction, observationalism, conventionalism, and the denial of the role of the concept of truth), and secondly by providing a theory of what can be known by observation — which will determine how much correspondence is needed for observations to correspond with reality.

This last attempt, however, reveals a threat from a different quarter: the influence of frameworks of thought upon our knowledge. In other words, the long-neglected influence of metaphysics upon science is thrust upon us. Not only are we concerned with the unfashionable question of what sorts of things there may be in the universe, but also whether they can be reasonably claimed to exist.

From further consideration of the criterion of refutability, it transpires that in science we deal not only with the content of theories, which tell us how things work in the world — regarded usually as the sole function of science — but also with the basic structure of the world, here called embedded ontology. The significant point that emerges here is that this is a form of metaphysics which is not susceptible to any kind of scientific proof.

Along with this transpires another ingredient — the weltanschauung of a scientific theory — seen not as an excrescence which competent successors will eliminate, but as part of the life-blood of science itself, determining what shall and what shall not count as science, fostering its growth or just as often impeding it. Here is another form of metaphysics, long dimly recognised though not sharply detached for scrutiny, underlying science. This additional form leads to a significant new question about the role of philosophy — or more specifically of weltanschauungen — in scientific progress, and about the unique difficulty of tackling this factor. It is pointed out that the difficulty hinges on its connection with creativity, with all the mysteriousness of this phenomenon; for to become self-conscious about the phenomenon is to take the first step towards coming to grips with it.

The last really large development in philosophy of science appeared in 1934 when Sir Karl Popper published his *Logik der Forschung*. Since then there has been nothing on that scale; there has been elaboration of what had gone before, and there has been controversy — the counter-culture has even found its way in. The present work touches on these matters only briefly, if sharply. This is partly because the book is concerned with a different problem: concerning a novel relation between science and metaphysics; and therefore it takes a wholly different direction. While it is not a polemic rallying to Sir Karl's side, it does take off from Popper's theory of testability and refutability. His theory, with relatively minor adjustments, I regard as acceptable so far as it goes. But as a philosophy of *science* it fails to map the whole area of science. Briefly put, this is because Sir Karl restricted himself to the testable components of science, and thus gave no treatment of the metaphysical underlay. *This is the most central feature of the present enquiry.*

It may help to explain the close relation of this work to Popper's, and its development away from his, if I recount its origin. Accepting his criterion of refutability as basic for determining whether a proposition belongs to science or not, I came across counter-examples. It soon became clear that there were indeed propositions, very strictly belonging to the corpus of science, that could not be refuted or tested in Popper's way, that they were inherent in and ineradicable from science, and that they were metaphysical. Accordingly, I sought a theory that would deal with their role, and some new method of refuting them, which would endow them with some semblance of rationality. And it then turned out that such propositions prove to be highly influential over scientific — and other — thinking.

It is worth mentioning a twofold overall objective. I have been long concerned with the problems discussed here, interested in them in their own right, and trying to find solutions, because of their significance for natural science, and for rationality of outlook in general. In addition, however, I have become impressed during the last decade with the enormous importance these issues have, indirectly but also directly, for the social sciences. This book, therefore, as well as attempting to contribute to the philosophy of science in general, is a prelude to a further work, on the social sciences.

But before we reach that, our basic question here is: what is the role of philosophy of science in doing natural science itself?

Notes

1 Although I am concerned especially with anti-essentialism in the origin of modern science, this does not deny the importance of other factors, such as the search for certainty (observation in Bacon, reason and mathematics in Descartes).
2 For a useful elucidation of widespread misunderstanding of Popper's philosophy of science, especially how we can have objective knowledge even though all knowledge is fallible, see Radknitzky (1982). I might mention — what is hard to come by — a serviceable elementary little book on the philosophy of science by Munévar (1981).

References

Munévar, Gonzalo (1981), *Radical Knowledge: a philosophical inquiry into the nature and limits of science*, Avebury, Amersham.

Popper, K.R. (1934), *Logik der Forschung: zur Erkenntnistheorie der modernen Naturwissenschaft*, Springen, Wien (imprint, 1935).

Popper, K.R. (1946), *The Open Society and its Enemies*, London and New York, Ch.3.

Popper, K.R. (1963), *Conjectures and Refutations*, Basic Books, London and New York, Introduction, 'On the Sources of Knowledge and of Ignorance'.

Radknitzky, Gerald (1982), 'Knowing and Guessing: if all knowledge is conjectural, can we then speak of cognitive process?', *Zeitschrift für allgemeine Wissenschaftstheorie, 13*, pp.110–21.

2 The origin of modern science: observationalism and induction contra essentialism

There are hard-headed experimentalists on the one hand and hard-headed theoretical scientists on the other hand. Both of these groups look on modern science as something that has become freed from its medieval origins. They might be willing to concede that in the Renaissance the scientific revolution retained survivals of medieval thinking, but they would hold that these blemishes disappeared in the course of time and have long been liquidated from modern science. It is certainly to be agreed that modern science has succeeded effectively in eliminating certain anti-scientific features of medieval thinking; but the thesis to be put forward here is that there still remain manifestations of the origins of modern science, and that these exercise a strong influence on science today. More specifically the contention is that the successful domains of science have not been interfered with at all, though in some cases their progress may have been slowed up by the reaction against medievalism, and that the possibilities of a much wider scientific development have been definitely impeded by this reaction.[1]

Still more specifically, the thesis is that human thinking was dominated almost exclusively, both in Greek times and in the middle ages, by essentialism (to give it the name introduced by Popper); that the rise of modern science consisted not so much in experimentalism but in rejecting essentialism, and finally, that the form that this took, making observation and mechanism ultimate, though not the only alternative available, was the most obvious one to follow, and that it has been the dominating influence in science ever since.

This may read as a strange thesis to practising scientists, even of the non-hard-headed variety who will be scarcely aware of the existence of the outlook dominated by essentialism, let alone the possibility that contemporary science is still mounted in opposition to it. The view suggested here, however, is that this historical situation is one of the most basic factors of all in explaining the existence and nature of modern science. Moreover, it is contended that it determines what scientists shall allow themselves to count as science. The investigation of all this alone provides an adequate raison d'être for the philosophy of science.

If science is built up upon a repudiation of the notion of essentialism, we have first to come to terms with this curious notion. The best account and criticism of the position was given by Popper (1946). To fix our ideas as rapidly as possible, we may recall that Darwin's theory of evolution rejected a notion of a fixed essence discriminating natural kinds of animals from one another and, in particular, human beings. According to his theory, it would be possible in principle to find the continuous shading or spectrum of animal kinds from the ape to man. What the theory repudiated was the notion of a sharp-cut distinction between animal and man. And without a sharp-cut discrimination no notion of a specific essence peculiar to each is possible.

The early history of the notion of essence is still of vital significance. Without going back to the earliest of the Greeks, it is a key notion in the philosophy of Plato, and his thinking has pervaded western civilisation for over 2,000 years. With Aristotle the notion assumes a particularly interesting form. It arose in the context of logic gracefully enough, where Aristotle, as Popper has made particularly clear, was concerned with different kinds of definition, the most important of which being what he called 'real' definition. This word is not very satisfactory, and Popper has rightly introduced the term 'essentialism' as more immediately informative: thus Aristotle was concerned with the theory of essential definition. The objective was to define the essence of some real thing such as a horse or a man. This was not a logical exercise but aimed to provide a clear account so that a discussion would be better understood. That this was very strikingly a philosophy of science is seen very simply by his use of it: intuitively it seemed obvious that we could define the essence of man by discriminating his genus and species by inner reflection — not by dirtying our hands by mingling with men in the market place. Aristotle's definition was that man is a 'rational animal'. It is easy to see that this provides factors in the essence without which an object could not be a man.

Our present concern is not with criticisms that might be levelled against Aristotle's definition but with the philosophy of science founded on it. Having arrived at his definition, he believed that by means of syllogistic logic it would be possible to derive consequential properties of man, and in this way the conclusions of logical inferences would provide us with all the most important knowledge about man. Thus Aristotle's essentialistic definition was a theory of scientific procedure. From the three parts, essential definition, logical deduction, and derived properties, one supposedly obtained knowledge about real entities in the world, and that constituted the notion of science.

This broad notion of essence and essentialism, whatever minor changes may have taken place, held for about 1,800 years. Nevertheless, it has not been unopposed: there were such figures down the ages as Antisthenes among the early Greeks, and William of Ockham in late medieval days. Nonetheless the mainstream was essentialistic.

The present thesis is that the scientific revolution was characterised by anti-essentialism. This was notably articulated — the idea was already working its way through in scientific theorising — by the philosopher Bacon in his *Novum Organum*.

The subsequent sporadic attacks on essentialism are especially interesting and worthy of notice. A very devastating one, carried out with clarity and thoroughness, was provided by Berkeley and repeated by Hume in the early eighteenth century. In the present century essentialism has come under a barrage of fire from very different kinds of philosophers. The logical analysis of philosophers headed by Russell and Moore, the logical positivists, the linguistic philosophers, all form one continuous growth rooted in anti-essentialism and constitute a worldwide attack. Along with this school is the attack made by Popper who is, in this respect, at one with them. Even stranger bedfellows are to be found in the ranks of the existentialists. Thus the most dominant streams of twentieth-century philosophy have all been anti-essentialist.

Returning now to Bacon, he was imbued with the need to escape from the myths, superstitions, unrealistic thinking, and irrelevancies of the middle ages. This was to be done, not by detailed sniping at one point after another, but by a wholesale sweeping away of the old with a new broom; and in this he reflected the attitude that animated the science of his time and before it, though not articulated explicitly. The question that arises is what was to be put in its place. After all, essentialism was a philosophy of science, the means of obtaining knowledge about the world; and if it was repudiated, there would be a vacuum. Yet alternative means of attaining knowledge could be developed. Bacon's idea was very simple and very sensible; and although it is, I think, quite wrong, at least one should understand its reasonableness in the situation. Since all the absurdities of the preceding era had arisen because of the belief in acquiring knowledge simply by reflection, the alternative was to go out into the world and explore. And what do you do when you try to find out about men or about things when you explore them? You begin by observing them. Thus it seems very natural that Bacon should have replaced pure thinking, pure reflection, by a method of observation. He realised, of course, that observation was not enough; observation supplies the raw material, but somehow one has to generate knowledge, or, alternatively expressed, one has to generate truth. So Bacon had to provide a method of generating truth out of observation. This was done by a new form of logic called induction. The function of induction was to arrive at general truth or generalisation from particular observations.

It is especially important to note the twin features of Bacon's new theory, which was also a philosophy of science. They are a philosophy of observationalism and a logic of induction.

This may not startle contemporary scientists, who mostly have been taught that science begins with observation and proceeds by induction. They may think that everything said so far in this respect is merely a repetition of an old story; but there are two reasons why this is not so. One is becoming fairly well known: that is, science does not begin with observations and does not use induction. The 400-year-old myth of science as observational and inductive has been one to which almost all the great physicists of the past subscribed even though, if they had bestowed a little attention on what they were doing, they could easily have seen that their practice in no way accorded with this view. This is now becoming recognised fairly widely; but it is by no means universally so, either among scientists or among philosophers. The other reason why this is not just a repetition of an old story lies in the following thesis: that *the myth of the philosophy of observationalism and the logic of induction* has persisted for so many centuries and *still persists because of the power of essentialism* which it superseded; because, in fact, *it draws its sustenance from opposition to essentialism*. It is reasonable to suppose that scientists and philosophers feel insecure when observationalism and induction are attacked because of the fear that this might lead to a restoration of essentialism, and nullify the gains of a very long and bitter struggle to get rid of it.

Bacon faced a very serious problem. Assuming that science cannot be conducted simply by reflection and clear thinking, he very naturally thought we must begin by making observations. There arises a quite fundamental question: how is it possible to make inferences from observations (or facts) to theories? The reason why this is a fundamental matter is that Aristotle's logic shows that the procedure cannot be done. There are great numbers of objections to induction, but here it will suffice to mention that from Aristotle's logic it is clear that if premisses are about particular things, and if the conclusion or theory or generalisation is general or universal, the inference must be invalid; or, in other words, a universal conclusion cannot validly be deduced from particular premisses. The social and psychological aspects of the situation are also very interesting at this point. The fallacy of deducing universal conclusions from particular premisses was a simple and obvious consequence of Aristotle's logic, and was not due to a defect in that logic. This has been recognised off and on for centuries. Many of the great medieval logicians would have recognised the impossibility of arriving at truth by this method. How then did Bacon and his successors in philosophy and science continue for 400 years not just to pay lip service to this philosophy of science, but actually to seek devious ways to obviate the difficulty? The need to believe it must have been very great, sustained in order to prevent a throwback to essentialism.

However that may be, the belief in the philosophy of observationalism and induction persisted from Bacon's time to the present century. It is indeed the most influential view still persisting in philosophical circles, though it has been eliminated in the more fundamental natural sciences such as physics. It persists quite noticeably in areas such as physiology. That is, it persists as a belief, though not necessarily as practice. Indeed, if Popper's philosophy of science, which is tending to replace it, is correct, as in large part it would seem to be, this shows not only that science has not been inductive but that in principle it cannot be so. Whatever attempt scientists may make to practise induction, this is something that cannot be done — this is not something that cannot be done well, for it cannot be done at all. Some sort of approximation to it can, however, be done and may carry a misleading air of similarity to it.

The opposition between essentialism (Platonism or Aristotelianism) and observationalism and induction (Baconian science) has long been recognised, and has been particularly emphasised by Popper; but the point being made here is that the dynamic of this opposition still persists.

Appendix: Known Objections to Induction

Overwhelming fallacies in induction are by now well known (Wisdom, 1952); some are undermining, others decisive. These and some other difficulties may be assembled briefly as follows — though I believe that even more fundamental objections exist.

Some years ago, I distinguished between induction as an inference and induction as a method (Wisdom, 1952). They had of course long been implicitly recognised to be distinct (notably by Mill), although the distinction was seldom followed up. It seemed to me that it made discussion and criticism more telling. It was Lakatos (1968), however, who made full use of the distinction, which he newly conceptualised as between justification and heuristic. Historically, justification was the more ultimate problem, heuristic the more ultimate aim. From Bacon onward, what was sought was justified discovery. It was assumed that justification would ensue from a proper method. Induction was intended to be an algorithm for computing relations between observations.

It has been denied that induction exists. It is, I think, true that inductive method does not exist. It is, I think, also true that inductive inference does not occur in science. But inductive inference certainly exists, for examples of it can be written down and are to be found in any traditional textbook of logic: any inference with a finite number of observational premisses and a generalisation as a conclusion is inductive. It is invalid — but it exists.

Method/Discovery

1 It is impossible in principle to pass by induction from generalisations to higher-level theories (involving unobservables, theoretical terms, or non-instantive concepts), for the terms involved are not the same in the theory as in the generalisations, as they must be in an inductive process. This point gains additional strength where the generalisations, such as Kepler's laws, supposedly subsumable under a theory, such as Newton's, turn out to be only approximations.
2 It is impossible to obtain observations for an induction without making a selection, that is, presupposing a hypothesis.
3 The inductive method, as conceived in the Bacon-Mill tradition, would turn scientific research, at least as regards generalisations, into an algorithm — a reductio ad absurdum of scientific research.

The situation has altered of recent times. The greatest contemporary inductivist, Carnap (1946, 1962), seems to have agreed that the method of science is hypothetico-deductive, making bold guesses at improbable hypotheses and testing them; in other words, that the traditional conception of inductive method has nothing to do with the method of science. This has an undermining effect on the conception of inductive inference, because the original aim of the method was to provide reliable results which could be regarded as conclusions of an inference. Still, the inference may be detached from its raison d'être and considered in its own right or provided with a new raison d'être (which, incidentally, has not been done).

Inference/Justification

4 An inductive inference commits the classical fallacy in formal logic of illicit process, where a term is particular in its premiss but universal in the conclusion. A counter-example can be found containing a true premiss and a false conclusion.
5 If induction were successful, it would prove too much; for, from correct observations, we should obtain a generalisation we could not

deny, which would not accord with the modifiability in principle of all scientific laws. This constitutes a reductio ad absurdum of inductive inference. As my former student, Mr Thomas K. Bowers, has pointed out, inductionists would have to hold that a new induction *must* be consistent with all existing inductions.

6 There is good reason for the contention that all scientific theories, including generalisations, are false. If so, inductive inference would establish what is impossible. This, too, would be a reductio ad absurdum.

A long line of inductive logicians (see Wisdom, 1952), culminating in Carnap, has, however, replaced the old goal of certain knowledge by the probabilistic goal of showing that a generalisation has its probability increased by additional observational confirmatory evidence.

7 Even the probabilistic inference cannot be got off the ground in an infinite, or even large, universe, without assuming some theory about its nature; in other words — though still controversial — it is hard to see how to obviate Popper's contention that the probability is and always must be zero.

8 Moreover, as Hume (1938) pointed out, a probabilistic argument from the present to the future presupposes a principle of probable connection between the present and the future.

Psychology/Philosophy of Mind

9 Scientific discovery, of the kind recognised as being one of the most highly imaginative activities of mankind, would be without a place in science. A training in skills of B+ level would suffice for research.

10 The conception of the human mind would be that of a tabula rasa, receiving impressions imprinted on it. Popper (1946) has updated Locke's conception as the 'bucket theory' of the mind, but more apt would be the 'stomach theory', for the data are ingested and digested by a mechanical process.

The above points are not precise refutations of induction as a method, but they are undermining in showing up its unreality.

Note

1 There are, of course, many factors influencing the development of science. I am grateful to my former student, Dr Michael Haynes, for emphasising the widespread demand for certainty, and also the deep influence of the model of Euclidean geometry.

References

Bacon, Francis (1620), *Novum Organum*.
Carnap, Rudolph (1946), 'Theory of Prediction in Science', *Science*, *104*, p.520.
Carnap, Rudolph (1962), *Logical Foundations of Probability*, University of Chicago Press, Chicago, pp.192–3.
Hume, David (1938), *An Abstract of a Treatise on Human Nature*, Cambridge University Press, Cambridge, p.15.
Lakatos, Imre (1968), 'Changes in the Problem of Inductive Logic', *London Colloquium on the Philosophy of Science*, North-Holland, Amsterdam, Volume 3, p.315.
Popper, K.R. (c.1946), Unpublished Lectures.
Thompson, Sir George (1963), 'Some Thoughts on the Scientific Method', *Boston Studies in the Philosophy of Science*, pp.81–92, ed. Cohen and Wartoffsky, Humanities Press, New York.
Wisdom, J.O. (1952), *Foundations of Inference in Natural Science*, Methuen, London.

3 Four contemporary interpretations of the nature of science

Introduction

If it should seem that there has been no rival to induction until Popper's theory of falsifiability, this impression must be corrected. Conventionalism, developed at the turn of the century, was a wholly new interpretation of the basic structure of science. It influenced scientific thinking during this century and even had affinities elsewhere (e.g. phenomenalism in philosophy). Moreover, an old doctrine, instrumentalism, was revived in the 1920s. This chapter will therefore concentrate on the four competing interpretations of science that have been in the foreground of philosophical enquiry in this century.

The topic of this chapter might sound as if it is merely the subject for a technical exercise: but it springs from a wider concern. In my opinion hardly anyone in this scientific age knows what the nature of science is, and this includes not only those who are aware of science only through weekly magazines, but also philosophers and scientists themselves. It may come as a surprise to most scientists to hear that there are competing views of the nature of science sponsored by exalted scientists. Most scientists who have had little contact with this question tend to assume (a) that the nature of science is obvious, (b) that there is only one possible answer for those who really know science, (c) that all good scientists are agreed or would agree about the answer, and (d) that in any case the serious researcher, while he may give his interested brethren the same tolerant smile he would give to any hobby, knows the whole matter has no bearing on the progress of science itself. This last, if wrong, could have enormous prejudicial effects. There is nowhere near universal agreement among the ablest scientists; hence, it is false that there is only one possible answer for those in the know; and hence the answer is not obvious.

There are surprisingly many different interpretations around. Apart from Popper's relatively recent addition, there have been for many years some seven more or less orthodox contemporary interpretations, though only four of these are significantly different from one another (moreover, these in turn are greatly different from those of two or three centuries ago, which themselves were somewhat different from one another, for example, those adopted by Galileo, Descartes, and Newton). Why is there such a discrepancy about something that should be straightforward? I presume there is some problem these interpretations are intended to solve. I suspect that they all stem from the underlying problem of justifying theories in the light of observations. Whatever their origin, it will be argued that they fail to solve this problem.

Three of these form a group, and they all come from scientific sources. They are instrumentalism (the most extreme), conventionalism, and operationalism. A complication that hampers discussion is that several different categories underlie these interpretations. Thus, one is concerned with concepts, the others with theories; two are concerned with meaning and reality, two with truth. The apparent disparity of aims makes it somewhat difficult to compare and contrast them. Such disparities, however, will fall into place to some extent if it should turn out that a common concern was the relation between observations and ideas, of which meaning, reality, and truth are all facets. While it may help the reader to bear this background in mind, I do not propose to develop the interpretations in terms of it, nor in terms of my conjecture of the underlying problem. Historically speaking, they probably have all sprung from the same root in twentieth-century physics; nonetheless, their originators were impressed by different facets of this science. Moreover, no one of these interpretations was developed in opposition to another. I shall therefore present each in terms of the immediate situation that gave rise to it.

I shall begin with these three: I hope to exhibit them in a new light which may make it easier to assess them. The fourth, induction, comes from a long-standing philosophical tradition, and I shall discuss it separately. In the end, I shall relate them to Popper's interpretation, which is a heterodox position becoming orthodox.

Instrumentalism

One highly influential interpretation arose — or was revived — because of a unique situation in quantum mechanics. Because of the uncertainty relations and the causal anomalies arising here, one could not, apparently, speak truly about various motions: all one could do was to calculate and predict. This gave rise to the doctrine that scientific theory is only a tool, only an instrument, for calculation or prediction. In brief, a theory is a computation-instrument. No one would deny that theories are at least tools; what the doctrine claims is that they are *only* tools.[1] The doctrine was aptly named 'instrumentalism' by Popper.[2]

The doctrine is very old: it existed with Berkeley[3] (1721) and in medieval days, and was at the root of the struggle between Cardinal Bellarmino and Galileo. In contemporary days, one may take Heisenberg as being for most of his life an instrumentalist, believing that the theories of quantum mechanics said nothing about realities, but were simply tools for prediction. Although the doctrine is both very old and very new, it came from totally different sources at different periods in history. Initially it sprang from a theological root. Centuries ago it was dogmatically held that the only really fundamental cause was God, the prime mover, and no other kinds of causation were independent; if one had ostensibly another kind of causation, it was secondary and did not, so to speak, have the punch of causation in it — one only had regularities on the surface. Thus, when one billiard ball knocked another one and, as we might say, caused it to move, this would be a description, according to the theory, which should not impute causal efficacy to the ball; the causal efficacy would reside in God. Quite clearly, the modern quantum physicist does not take this line; his reason for adopting instrumentalism arises from the difficulties that lie in quantum physics itself. However, I am not concerned especially with these origins; I only want to make the point that instrumentalism does exist today in a new form. Whenever it has appeared, it has been very powerful, and today it is one of the most influential ways of looking at science. Instrumentalism is uninterested in the question of reality. It is inspired by logical positivism. It is positivistic in the sense that it is initially concerned to deny a reality underlying observations. But it is more extreme, for logical positivism was at least concerned with truth, whereas instrumentalism dispenses altogether with the concept of truth. Its raison d'être lies in the suspicion that any alternative is metaphysical.[4]

Mathematicism At this point it is necessary to refer to a view of science that characterises it as being essentially mathematical. One constantly finds people saying that to be scientific a discipline must be mathematical. It is not easy to document the view, but perhaps Jeans (1931) would have subscribed to it. It is not the sort of doctrine that is thought out or written about, and since it is not well articulated, it is not so easy to characterise. But it means at least that science is no good unless it is mathematical. There surely is something in this, insofar as it prizes accuracy and can lead to discoveries not obtainable otherwise, but it is also surely off the mark since, as has been clearly brought out by Webb (1944), much good science is not mathematical. But what I think is meant also is that it emphasises the ability to calculate. If this is the nub of the claim, it amounts to a variant of instrumentalism. I doubt if instrumentalism is its only component, but instrumentalism would seem to be the most important one; at any rate, it seems to offer nothing clear and distinctive beyond instrumentalism.

Moreover, in Galileo's time, what Popper calls the instrumentalist view of science was manifested officially by speaking of 'mathematical hypotheses'.

Instrumentalism is still the establishment view in physics, though critics like Landé are trying to stem the tide.

It seems to be widely overlooked that instrumentalism is the representative of pragmatism in the philosophy of science. There may be no explicit pragmatist philosophy of science as such, and probably no contemporary philosopher would call himself a pragmatist (in the sense, for example, of subscribing to its theory of truth); but pragmatism informs almost every philosophy of science of the century — simply through the broad approach that an idea has to work. The linkage between instrumentalism and pragmatism is somewhat closer, in that both share a lack of interest in the reality behind what works: both in effect deny meaning to concepts in order to obtain statements that work.

Operationalism

Operationalism will be interpreted as a variant, but it has had such prominence that it first should be considered in its own right. It began in America in 1927 with the eminent physicist Bridgman. The doctrine is that a concept has no meaning in itself; its meaning is what one does with it: a concept does not designate, but gains a meaning from the operations one carries out when one uses it. Thus, the operational definition of 'length' is given by operations of measuring. (I doubt if it is possible to formulate this doctrine properly, in particular in a non-circular way; but exactitude about formulation is not necessary for my purpose.) Since Bridgman was apparently very impressed by the problem of whether there is a reality answering to certain terms in the special theory of relativity, he took his main example from there. Thus, he held that there is no meaning of length in itself, and no such thing as length in nature: but there is the operation of measuring, and when certain operations of measuring come out with roughly the same answer, then we have a quantity we can call 'length' defined in terms of these operations. The great difficulty with operationalism, apart from formulation, is to make it work at all plausibly. It has some plausibility with 'length', but almost none with very abstract concepts such as 'electromagnetic field'. Bridgman (1927) faced this and tried to meet it by including paper-and-pencil operations. Operationalism is overtly concerned with concepts, not with theories; with a criterion for meaningfulness of concepts; and with a sophisticated conception of reality. Operationalism should be closely linked with logical positivism. Logical positivism has been concerned to deny reality behind observations, and to provide a criterion for the meaning of concepts. In both these respects it resembles operationalism, but it differs in having as its main focus a criterion for the meaning of theories. Thus, it is a somewhat broader view. Since, however, it concentrates almost exclusively on the question of meaning, it hardly amounts to an interpretation of science — it tells us whether a piece of science is meaningful or not, and that is all. However, there is no question that the spirit of logical positivism has permeated a good deal of scientific thought, and thus influenced the interpretation of science held by many eminent physicists such as Bridgman.

Operationalism has largely disappeared from the outlook of physicists, though it still permeates the thinking of many social scientists. Broadly speaking, it would seem that the raison d'être of operationalism is suspicion about metaphysical reality.

Is truth-value involved? The reason for labouring this is that it is important to decide whether these interpretations are in each case concerned fundamentally with truth-value, or whether they are not.

Operationalism, not being concerned with theories, is not concerned with truth. It is concerned with meaning and reality; and here the aim is to deny hidden reality and affirm positive reality or reality that may be confronted in observation. We could add, however, that operationalism is a form of conventionalism (restricted to concepts) *at the observation level*. That is to say, the meaning attached to the concept of 'length' is a convention adopted to render coherent the operations of measuring. Even so, however, the doctrine does not go so far as to make the meaning of operations conventional. (It is characteristic of all these interpretations of science that it is impossible, difficult, or at least implausible to push them to the bitter end.)

On the other hand, operationalism is not concerned to deny truth-value to theories. So far as it might do so, it would imply a conventionalist interpretation. Thus, operationalism extended broadly enough to be an interpretation of science would be indistinguishable from conventionalism. There is therefore no further need to consider operationalism as a serious rival interpretation of science.

Conventionalism

It has become customary in philosophical circles to attribute conventionalism to Duhem (1962, orig. 1904) when discussing it in relation to recent revivers of the doctrine, notably Quine (1953), Kuhn (1962), and possibly in effect Feyerabend (1962). Now Duhem's chief contribution was probably his emphasis on the fact that when a test turns out to be false and constitutes a refutation, it refutes not a single hypothesis or theory, but the entire theoretical structure of hypotheses, initial conditions, and so on, from which the test observation is deduced; so that there is no means of knowing which ingredient has been falsified.[5] This highly significant conclusion was, of course, taken over by Popper (1959), although he absorbed it in his realist interpretation of the nature of science. But Duhem in effect took it over from Poincaré (1952, orig. 1902) who, though he may not have articulated it quite so overtly as Duhem, did make it virtually explicit. Poincaré (1958) himself had some difficulty about conventionalism, but I think it would be true to say that he held it, though with reservations. His reservations had lead him to criticise very severely the philosopher of the day who went the whole hog with the conventionalist interpretation — le Roy[6] (1900). Nonetheless, I think the conventionalist interpretation should be called the le Roy-Poincaré thesis.[7] Among discussions explicitly focused on conventionalism, I have come across a useful one by Nancy Tuana showing that Duhem, contrary to general belief, was not a conventionalist; unfortunately I cannot find a reference.

To illustrate more concretely, if in geometry we alter the conventional meaning of the term 'straight line', we obtain a new system which is just as valid as the initial one. Thus, Riemannian geometry is just as valid, as a system, as is Euclidean geometry, but in it the term 'straight line' assumes the new meaning of 'geodesic', which is the shortest distance between two points, though it is not 'straight' in the ordinary Euclidean sense (thus a great circle on a globe is a geodesic). The central idea here is that one theory may be exchanged for another one by altering the conventions to do with the meaning of basic terms. And here we have the central idea of 'truth as a convention', not as recording something about the world.[8] Expressed conveniently though metaphorically, a theory can be regarded as a pattern, drawn not because it reflects reality but because it brings order of some sort into our observations; and this order or pattern is conventional, because, if it does not produce adequate order, we can modify the pattern to make it usable. In effect, theoretical terms on this view have no denotation. (This has been neatly formulated by Agassi, 1966, pp.4—5.) The relationship between framing conventions and ordering our observations is that order is the basic objective subserved by manipulating the conventions suitably. Conversely, conventionalism renders a theory irrefutable, that is, irrefutable by observation; for even if a theory does not square with its predictions and if for the time being no alternative convention is available that would make it square with them, nonetheless there is always the possibility that a new convention might be thought of that would save the theory. But if irrefutable, neither is a theory conclusively verifiable: there always exists the possibility of putting together an alternative theory that would equally well explain the phenomena covered by the given theory and lead to the same predictions as the first one; but since no theory is refutable, the alternative theory could not be refuted, and, so long as this is the case, the given theory could not be regarded as conclusively verified. Thus acceptance of a theory would be conventional because it would be an arbitrary choice of one out of a set of possible theories that would do the same work. The disturbing feature of arbitrariness is metascientific in that observations are not decisive; the practical criterion would be practical convenience.

Thus, a theory cannot be accepted or rejected on the grounds of being true or false in the light of evidence (for the evidence is not conclusive even though it is all we have); it can be maintained or retained simply by adjustment of the meaning of a key term, that is, by changing the conventions for the use of a term.

Putting these facets together we have the following: First, conventionalism starts intellectually from the irrefutability of theories in an absolute sense, since a discordant observation refutes a system as a whole and not the theory alone. Second, dwelling on this we come to its consequences: that alternatives are also irrefutable, and that no theory is verifiable; in short the truth-value of theories cannot be absolutely settled by evidence. However, though refutation is not absolute, evidence is sometimes interpreted as being against a theory. Third, all this fits agreeably into a setting in which it is recognised that the meanings of terms can be adjusted by convention, and adjusted in such a way as to preserve the theory against the discordant evidence, that is, to preserve its truth (the realistic difficulties of refutation encountered by Poincaré fell on a suitable soil, the impact of alternative geometries).

Irrefutability alone, without making the third feature explicit, is not sufficient to constitute conventionalism. All that it asserts is that, since we cannot possess conclusive evidence, therefore we do not have absolute knowledge, and therefore we do not have *knowledge* — only conventions or 'as if' theories, that is, our observations cohere as if our theories record something about the world. Such a conclusion may give the erroneous impression of being very cautious, since, in the absence of any basis for the truth of a theory, there can be no such thing as a scientific description of the world even in principle. In fact, this sweeping negative conclusion is the reverse of caution. All that is warranted is that we can never tell whether or not our theories describe the world, and this is Popper's (1963) line: a scientific theory may very well be absolutely true but we can never know it. If this modest agnosticism were all that was being claimed by conventionalism, then conventionalism would hardly be an outstanding interpretation of the nature of science. Let us therefore turn to the other ground for the doctrine. It is true that Poincaré was overwhelmingly impressed by the difficulty of refutation, and quite rightly so, but I would suggest that this only *raised* for Poincaré the problem of the truth of scientific theories. Poincaré did not erect conventionalism on this basis alone; and in support of this last point it must be borne in mind that it is impossible to think of Poincaré except against a background of the great modern developments of geometry and algebra, not forgetting that he was himself a contributor to this geometry. The decisive factor would seem to be the consistency or, if we will, the coherence of these modern geometries, which are just as consistent as Euclidean geometry, with no way of choosing between them — it is this that provides the raison d'être of conventionalism.

In short, the impossibility of pinning falsity on a theory because it may lie anywhere in the whole theoretical system containing the theory paves the way to the real basis of conventionalism, namely that although a theory is irrefutable in principle, it is threatened, but the threat can always be removed and the theory saved by altering the conventions.

In this background we may notice the confluence of *irrefutability*, *equivalidity of theories*, and *meaning by convention*. Irrefutability lies at the root of equivalidity, and equivalidity turns on the possibility of changing the conventions — or so, at least, it would look to anyone immersed in four-dimensional geometry, for the different systems of geometry may be interpreted as a difference of convention about the meaning of the term 'straight line'. If we extend this notion of conventionalism to the special theory of relativity, we get the result, not that the three-dimensional metric of Euclidean geometry is *false* and that Einstein's four-dimensional metric is *true*, but that the three-dimensional *convention* yielding the metric has to be *replaced* by the four-dimensional *convention*.[9] In short, it is not that there are different basic forms of conventionalism, but that it contains a tiered structure of equivalidity, irrefutability, and conventional meaning, each important in some contexts, the last being metascientifically basic.

There would seem, however, to be a deep-seated oversight in this approach. The device of convention-replacement will, of course, enable us to retain certain parts of a theory. Thus 'A straight line is the shortest distance between two points' remains true in four-dimensional geometry as it is in Euclidean geometry. That is to say, certain relatively *unspecific* propositions may be retained. But if we look to the more *specific* propositions that come under these, which are after all characteristic of the various theories, then these are simply not retained and have no equivalence at all in alternative theories, not even by convention. (Poincaré showed his concern about this in his discussion of the principle of conservation of energy.) Thus, the Galilean metric as measured in a moving frame is not the same as the Einstein metric, and the scientific position is first that both cannot be true and second that the evidence is against the Galilean metric.

Since conventionalism, even in the more refined Poincaré version, has something of the air of being a contortion, the question arises as to what is the underlying leitmotif for such an approach.

The conventionalist ploy worked beautifully in some areas. There are many such processes that worked very well in mathematics and logic, but which do not seem applicable in empirical science. Be that as it may, it seems to me that this was the underlying factor that provided Poincaré with a general approach to knowledge. Taking this line got him out of the difficulty of saying that a theory was an absolute truth. But we are still faced with the difficulty of understanding why this was a burning problem. The reason is that Poincaré was imbued with the difficulty of getting decisive evidence for or against anything in science. Otherwise expressed, he had an intuitive sense of difficulty about a realist interpretation. (There is the impact of the impossibility of establishing an absolute refutation; but, as indicated, this does not basically rule out a realist approach.) So, if one could not get truth, then the next best thing is a satisfactory convention. That was the leitmotif. The problem here has come out into the open in a virulent form, which will be the subject of a later chapter.

However, if one looks at conventionalism more closely, one sees that it introduces the notion of a certain kind of truth, though truth in a Pickwickian sense. Indeed, we should note that Vaihainger's philosophy of 'as if' is but a form of conventionalism. The peculiar sense of truth involved is the point at which difficulties begin to arise. Theories, for Poincaré, flatly had *no* truth-value; for they were true by convention. Thus, he was introducing a more sophisticated notion of truth than the prevailing one (which was of a realistic character or was a correspondence notion); it could be said to be a form of coherence theory. In short, one was concerned in the field of science with truth in some sense, but not the ordinary one.

Evidently, conventionalism stresses theories; it is certainly concerned with the meaning of terms, but only so far as they form part of a theory or contribute to its truth or falsity. Reality is a secondary consideration. Conventionalism does not deny reality but is, so to speak, agnostic about it: that is to say, all one can do is to make usable conventions about terms. Its primary concern is pattern-making. A highly scholarly discussion of Poincaré has been published by O'Gorman (1977).

Induction

The great traditional interpretation of science was, of course, induction. It was an invention of philosophers, but, remarkably, scientists mostly acquiesced in this philosophical characterisation, not noticing that their practice in no way accorded with it. Induction is the interpretation that science begins by making observations upon which is built, or out of which is obtained, by some form of reasoning, a generalisation. Theories, as more complex or more general or abstract than generalisations, received but scant notice, being regarded as no different in principle, but simply a further induction at a new level erected upon the initial generalisation or generalisations. The salient feature of generalisations or theories was that, if carefully processed, they were true. Induction was therefore an interpretation of scientific theories as having a truth-value. Induction is a method for arriving at truth and justifying the truth arrived at. And if a theory is true, it says something true about the world, about reality. Induction was also a method for making true predictions.

What induction does not accommodate is false theories (which is quite a gap, especially if all theories are false!). And of course it has, as is well known, a whole crop of internal difficulties (listed in the appendix to Chapter 2). It is simply invalid and can be rescued, if at all, only by unbelievable contortions. Here I wish to mention some cognates, because the general point at stake in this interpretation is, I think, wider than appears from the interpretation itself.

Experimentalism There is a theory of science maintaining that science must be experimental. Although one hears scientists saying this kind of thing, it has never been elaborated into a theory, so discussion of it is hampered. My surmise about what it means is that one should eschew a priori speculations: it means no stuff and nonsense, hardbaked realism, getting down to real work. This is all very worthy, but hardly adds up to a metatheory. And it is naive, since there is a great deal of science that contains no experiments at all.

Factualism In the social sciences, Watson, the founder of behaviourism, had no use for anything but facts. Moreover, he sought facts in two different places. He held that one could not study any object scientifically unless it was factual; thus, thinking could not be studied, but subvocal talking could. He also held that every ingredient of a *theory* has to be of a factual kind. This latter extreme has almost no following among natural scientists, who are well aware that an electromagnetic wave, attractive force, and things like that are unobservables. However, among social scientists, many — too many — have accepted the idea that the terms of a theory must be factual. And they would also tend to agree that the phenomenon to be studied must be factual.

It seems to me there is a spirit abroad that takes factualism, in one or both forms, very seriously. And perhaps it is not an accident that the social sciences seem to express this spirit. Great numbers of social scientists spend their lives trying to collect facts, and provide theories that are nothing more than a compendium of facts. In the field of the natural sciences, we meet the same view with those who hold that what really counts is pointer-readings. This powerful and widespread phenomenon is hardly articulated as such, but I am concerned with the metascientific views that are at work among our highpowered scientists whether these views are explicitly formulated or not, and experimentalism and factualism fall into this category. Further, it seems clear that experimentalism and the Watsonian type of factual behaviourism are forms of induction, and that they all contain one and the same hard core, namely the ultimacy of *observation*.

Resulting interpretations

It is worth noting that these approaches have come from eminent scientists in one of the greatest ages of science, the first quarter of the twentieth century: instrumentalism (Heisenberg, Bohr, etc); conventionalism (Poincaré); operationalism (Bridgman); and the minor ones, mathematicism (Jeans and many others) and experimentation (widespread, silent majority). Induction and observationalism are traditional, coming from philosophers with backbench support from scientists.

If we contrast the twentieth century with that other great age of science dominated by the big five, Copernicus, Kepler, Galileo, Descartes, and Newton, we notice one enormous strategic difference. Built into the twentieth-century scientific outlook is the belief that science is fallible. But for sixteenth- and seventeenth-century science, it could be certain. In significant detail its exponents differed. Galileo looked on experiment as a means of 'demonstrating', not proving or testing, but 'showing' what he knew to be true. For Descartes, falsification was a means of ruling out all but one of the possible theories, and the remaining one would be the truth. Newton alone regarded experiment as a means of testing hypotheses; but if tests were successful, the theory was considered established.

Of our seven interpretations, some have been subsumed under others. Mathematicism was included under instrumentalism. Operationalism can be treated as a special case of conventionalism. Experimentalism and factualism I have treated as variants of induction. Pragmatism and logical positivism were not taken to be specific interpretations of science. Thus, the serious contenders from the past are instrumentalism, conventionalism, and induction — with Popper's realism from the present.[10] It is worth noting, however, that instrumentalism is a particular case of conventionalism, in which truth-value in a Pickwickian sense is reduced to zero. Induction ought by now to be regarded as out of the running; instrumentalism is weak and we shall find reason below to think it can be knocked out; which leaves, or will leave, conventionalism versus realism as the living issue.

This issue is fundamental, first because of its extreme difficulty, but even more importantly because it raises the metaphysical question of the make-up of the universe.

Appendix: Meaninglessness in Quantum Mechanics

It has been a commonplace for many decades for physicists to speak of certain ideas, namely ideas from classical mechanics, as being *meaningless* in quantum mechanics. This rests on the confusion arising from the misapplication by physicists of logical positivism, though it should be added that some logical positivists themselves have not been free of the confusion.

The most central example concerns the 'uncertainty relations' which assert the impossibility of ascertaining, even in principle, the precise position and precise momentum of a particle. The pseudo-corollary has then been drawn that to speak of the precise position and precise momentum of a particle at the same moment is meaningless.

I will first illustrate by means of a simple example why this is fallacious. In non-Euclidean geometry it is an elementary theorem that the angles of a triangle add up to a quantity differing somewhat from 180 degrees. Is the assertion from Euclidean geometry that the angles make 180 degrees then meaningless in non-Euclidean geometry? It has long been recognised as a principle in logical positivism and also in its successor, linguistic philosophy, that if a statement is meaningless so is its negative, and if a statement is meaningful then so is its negative. In non-Euclidean geometry the assertion that the angles of a triangle are not equal to 180 degrees is true and if so meaningful. Hence in accordance with the principle so is the negative of this statement meaningful, that is, that the angles *are* equal to 180 degrees — it is just false. In short, the theorems of Euclidean geometry, given the axioms of non-Euclidean geometry, are not meaningless but simply false. Otherwise expressed, the theorems of Euclidean geometry can be disproved within the forms of calculi of non-Euclidean geometries.

The carry over to the parallel position in physics hardly needs to be given but in brief it is this. One of Heisenberg's uncertainty relations is that

$$\delta q \cdot \delta p \geq h/2\pi$$

Hence the contradictory of this, that

$$\delta q \cdot \delta p = 0$$

(which is true in classical physics) must be false in quantum mechanics; but nonetheless being the contradictory of a meaningful statement it must itself be meaningful (in quantum mechanics). In other words, in quantum mechanics, the assertion of precise position and precise momentum is false but not meaningless.

The error involved has, I believe, been very influential. In particular it has, I think, underlain the persistence of the instrumentalist account of science. On the other hand the converse may also be true, that the dominant influence of pragmatism and of logical positivism during this century would encourage exuberance about such dismissals and thus help to foster this erroneous interpretation.

Notes

1 What is possibly a variant was put forward by Mill (1843). He held that a theory was a principle in accordance with which one particular observation could be deduced from another particular observation; this has been revived by Ryle (1949), a theory being an 'inference-license'.
2 Apart from Popper's own, one of the four full discussions of instrumentalism is to be found in Agassi (1975).
3 Dewey used the term in an allied sense, but perhaps slightly different, in that he associated with it the pragmatic conception of truth.
4 In the paper on which this chapter is based, I associated pragmatism with conventionalism. Conventionalism and instrumentalism are very close, but I have come to the conclusion that, for all its instrumentalist overtones, conventionalism tries to take the concept of truth seriously; while for pragmatism truth is expendable. Hence pragmatism is to be identified with instrumentalism.
5 Grünbaum (1966, 1969) has argued in detail that it is possible in principle to pin the falsity firmly on the theory, but it is not clear how we can know a priori that no surprise will turn up affecting the initial conditions and saving the theory. In any case he (Grünbaum, 1969,

pp.1089, 1092) has granted that what he establishes is a 'well-nigh' conclusive result, that we can obtain falsification of a theory 'to all intents and purposes of the scientific enterprise' 'although we cannot falsify . . . beyond any and all possibility of subsequent rehabilitation'. This was never in doubt, and does not help to dispel the metascientific force of conventionalism which takes off from the unfalsifiability of a scientific theory in *principle*.

6   I am indebted to Mr T.C. Slate, when a student at the University of Southern California in 1965, for drawing my attention to le Roy.

7   My colleague, Professor Hattiangadi, considers it possible that there may even have been earlier conventionalists in the nineteenth century.

8   In other branches we get the same thing, for example, with the development of Hamilton's non-commutative algebra. A very simple illustration from elementary algebra is to be found in fractional indices; thus, $n^{1/2}$ has no meaning until it is given one by convention (which is done by making it obey the same rules as ordinary indices follow).

9   Interestingly enough, Professor Hattiangadi has pointed out to me, Poincaré opted not to change his conventions until Einstein's theory came out.

10   My student, Dr Malcolm MacKinnon, made the interesting point that instrumentalism, conventionalism and Popperianism of contemporary philosophy of science closely reflect the 2,000-year old dispute between nominalism, conception, and realism.

References

Agassi, Joseph (1966), 'Sensationalism', *Mind*, 75, pp.1—24.
Agassi, Joseph (1975), 'The Future of Berkeley's Instrumentalism', *Inter. Stud. in Philos.*, 7, pp.167—78.
Berkeley, George (1948—57), *De Motu* (1721), esp.§17, *Works*, Vol.4, ed. A.A. Luce and T.E. Jessop, Nelson, Edinburgh.
Bondi, Hermann and Kilmister, C.W. (1959), 'The Impact of Logik der Forschung', *Brit. J. Phil. Sci.*, 10, pp.55—7.
Bridgman, P.W. (1927), *The Logic of Modern Physics*, Macmillan, New York.
Dewey, John (1903), *Studies in Logical Theory*, University of Chicago Press, Chicago, pp.78—9.
Dewey, John (1963), *Philosophy and Civilization*, Capricorn, New York, pp.26—7.
Duhem, Pierre (1962), *The Aim and Structure of Physical Theory*, Atheneum Publishers, New York.

Feyerabend, Paul (1962), 'Explanation, Reduction, and Empiricism', *Minnesota Studies in the Philosophy of Science*, Vol.3, ed. H. Feigl and G. Maxwell, University of Minnesota Press, Minneapolis, pp.28–97.

George, W.H. (1936), *The Scientist in Action*, Williams and Norgate, London.

Grünbaum, Adolf (1966), 'The Falsifiability of a Component of a Theoretical System', ed. P. Feyerabend and G. Maxwell, *Mind, Matter and Method*, University of Minnesota Press, Minneapolis, pp.273–305.

Grünbaum, Adolf (1969), 'Can we Ascertain the Falsity of a Scientific Hypothesis?', *Studium Generale, 22*, pp.1061–93.

Heisenberg, Werner (1948), 'Der Begriff "abgeschlossene Theorie"', *Der modernen Naturwissenschaft, Dialectica 2*, p.333.

Jeans, James (1931), *The Mysterious Universe*, Macmillan, New York.

Körner, Stephan (1964), 'Deductive Unification and Idealisation', *British Journal of Philosophy of Science, 14*, pp.274–84.

Kuhn, T.S. (1962), *The Structure of Scientific Revolutions*, University of Chicago Press, Chicago.

Landé, Alfred (1965), *New Foundations of Quantum Mechanics*, Cambridge University Press, Cambridge.

Le Roy, E. (1900), 'Science et philosophie', *Revue de métaphysique et de morale, 8*, pp.37–72.

Medawar, P.B. (1963), 'Is the Scientific Paper a Fraud?', *The Listener, 70*, No.1798, 12 September, pp.377–8.

Medawar, P.B. (1967), *The Art of the Soluble*, Methuen, London.

Mill, J.S. (1843), *A System of Logic*, Parker, London, Book 2, Chapter 3.

O'Gorman, F.P. (1977), 'Poincaré's Conventionalism of Applied Geometry', *Stud. Hist. & Philos. Sc., 8*, pp.303–40.

Poincaré, Henri (1952), *Science and Hypothesis*, Dover, New York.

Poincaré, Henri (1958), *The Value of Science*, Dover, New York, Part 3.

Popper, K.R. (1959), *The Logic of Scientific Discovery*, Basic Books, London and New York.

Popper, K.R. (1963), 'Three Views Concerning Human Knowledge', in *Conjectures and Refutations*, Basic Books, London and New York.

Quine, W.V.O. (1953), 'Two Dogmas of Empiricism', *From a Logical Point of View*, Harvard University Press, Cambridge, Mass.

Ryle, Gilbert (1949), *The Concept of Mind*, Hutchinson, London, Chapter 5, pp.121 ff.

de Santillana, Giorgio (1955), *The Crime of Galileo*, University of Chicago Press, Chicago, pp.238–9.

Tarski, Alfred (1956), 'The Concept of Truth in Formalized Languages', *Logic, Semantics, Metamathematics*, Oxford University Press, New York, pp.152–278.

Vaihainger, H. (1924), *The Philosophy of 'As If'*, Routledge and Kegan Paul, London.

Watson, J.B. (1913), 'Psychology as the Behaviourist views it', *Psychol. Rev., 20*, p.158.

Watson, J.B. (1914), *Behaviour: An Introduction to Comparative Psychology*, H. Holt, New York.

Webb, D.A. (1944), 'The Place of Mathematics in Scientific Method', *Hermathena, 64*, pp.64–87.

4 The role of 'truth' in science

In this chapter I adduce a criticism centring on the role of truth-value, which constitutes an additional and powerful argument against instrumentalism and conventionalism, and therefore supports realism.

Instrumentalism and conventionalism relate differently to truth and reality. Since instrumentalism regards a scientific theory as being an instrument and *only* an instrument for computation, scientific theories are neither true nor false, they are only more efficient or less efficient instruments. Moreover, a theory has no power to describe the world of fact; suspicion about the question of reality or anything suggestive of metaphysics is the raison d'être of instrumentalism.

Conventionalism on the other hand allows of truth-value, though only in a Pickwickian sense, such that the conventions assigning the meanings of the terms of a theory are more or less appropriate. Conventionalism, however, though it holds out no hope of describing the world, adopts this attitude not because of suspicion of metaphysics but because of the apparent impossibility in principle of establishing either the truth or the falsity of a theory — and this very real difficulty is its raison d'être.

We thus have two types of view. One repudiates the notion of truth-value. The other interprets truth by means of a device. Neither accepts the ordinary notion of truth according to which a true theory is one that says something about the world. (The situation would be radically altered if it ever turned out that the ordinary notion of truth presupposed the conventionalist one at some level.) We may note, however, that instrumentalism amounts to a particular case of conventionalism in which the Pickwickian concept of truth-value is reduced to zero.

Turning to conventionalism, the position is that a deduced prediction would be false in an everyday sense, whereas the theory would be false in the sense of embodying inadequate conventions.[1] So the same argument would apply to a view that interpreted a theory as true by the use of a device as would apply to an interpretation that requires no concept of truth at all.

Let us elaborate the relation of truth-value or of the kind of concept of truth required for scientific inferences. The above argument amounts to this: refutation involves the idea that a conclusion deduced from a theory is false; this presupposes the principle of logic according to which in a valid inference falsity is transmitted from conclusion to premiss; and this presupposes that the concepts of truth and falsity applicable to a conclusion have to be applicable to a theory.

How far must we be able to specify the concept of truth or falsity required? Although I think that Tarski's definition suffices, I do not want to anchor the present thesis to his theory unnecessarily; and an important question may be raised about how much work it does in the present context. So I would put the matter in general terms thus. Those who would support, say, instrumentalism would have to deny to deduced conclusions truth-value in this sort of sense; otherwise a theory-instrument would have to have a truth-value in the same sense. On the other hand, to give up a theory-instrument as being a poor instrument rather than as false would have repercussions for the deduced conclusion; it could not have a truth-value. We could say only, 'The conclusion does not square with the actual state of affairs it would have squared with had it been a good instrument' (which would seem to be only a predicate replaceable by the shorthand 'false' in the ordinary sense).[2]

Conventionalism is in a better position. There is much to be said for it as a philosophy of science. It corresponds with scientific practice in some detail. Its great strength lies in two features to do with testability. It accords with the absolute impossibility of confirming a theory. And it accords with the enormous difficulty of refuting a theory. A theory cannot be refuted absolutely. For a theory to be rejected, we have to be satisfied that no loophole for saving it can be found in the surrounding premisses/assumptions involved in testing it. This cannot apparently be completely assured in principle, and though Grünbaum has made a strong attempt to deal with it, Poincaré was rightly impressed with this difficulty.

However, a grave difficulty arises over truth-value: conventionalism does not allow for the possibility of a large theoretical part for the concept of truth to play in a straightforward sense in the interpretation of science, even though in practice we may never be able to tell when we have a true or even a false theory; and conventionalism gives the strong impression that truth in anything like a literal sense is irrelevant to science; truth in a new, conventional sense is supposed to suffice. However, there is reason to maintain that it presupposes truth in the ordinary sense: for, as must be emphasised, in order to reject the conventions forming a theory, an observational conclusion must be false, and it must be false in the ordinary and not in the conventionalist sense. For, if a conclusion is false in the ordinary sense, then the theory it rules out must be false in the same sense, and not simply an inadequate convention.

Insofar as conventionalism relies on truth-value for a theory in a Pickwickian sense, it must do so for conclusions deduced from it. This extreme commitment has not been followed up by exponents of the doctrines.

Conventionalism could conceivably be remedied: once it is granted that a premiss must be false if the conclusion is false, then we could set about altering the conventions in the premiss. But then the conclusion could not have a truth-value in the ordinary sense.

The only way for conventionalism to win out — and this has not been seriously tried — would be to admit the role of truth-value and to show that it could play this role in the conventionalist sense, that is, that it could satisfy the requirements of the ordinary notion, that the conventionalist sense of truth could satisfactorily replace the ordinary one. The price would be that deduced observational conclusion would have truth-value in a purely conventionalist sense.

The overall point here is this: that the views discussed either lack the concept of truth-value required for scientific inference or use it in a sense that cannot be consistently pursued so long as test observations are given a literal truth-value. Thus refutation of a theory (regarded as an instrument), as a result of a conclusion that is false, is impossible, because an instrument cannot be false, only faulty. And refutation of a theory (regarded as a convention), as a result of a conclusion that is false, is impossible, because a convention cannot be false, only inappropriate.

Induction is not open to the same charge as instrumentalism and conventionalism, of not being able to accommodate truth-value. With all its (numerous) imperfections, induction has to its credit one great merit (not that this could save it) — it is concerned above all with truth. Induction is a procedure for obtaining generalisations from observations. The salient feature of an induction is that if the observations are true and accurately processed, the generalisation is *true* (or probably true). Thus, induction is a procedure that ascribes truth-value and tells you something true (or probable) about the world. Induction is nothing if not a way of arriving at truth and justifying the truth arrived at.

An important hard core of all the positions considered here is the ultimacy of observations, in the sense that observations are data and are certain, and that theories are of subordinate import.

We may note a small step of progress, for the argument of this chapter completely rules out instrumentalism. We are thus reduced to three possible philosophies of science: observationalism, conventionalism, and Popper's falsifiability. Then will come, according to these interpretations, the problem of what is real.

Notes

1 This inconsistency has also been noticed by Watkins (1958).
2 It will be recalled that conventionalists see science in terms of the hypothetico-deductive framework, in which an observational deduction functions as a test.

References

Watkins, J.W.N. (1958), 'Confirmable and Influential Metaphysics', *Mind,* 67, p.348.

5 The common problem: the observation – theory gap

The thesis I wish to put forward is that the three interpretations of science — inductive, instrumentalist, conventionalist — reflect different ways of handling observations. They are all concerned with knowledge derived from observations: scientists, it is presupposed, make observations and then have to put them together into a compound called knowledge — or more specifically, scientific hypotheses and theories. 'Nihil in intellectu quod non prius in sensu', which may be rendered: 'In the beginning was the datum, the observation, and without it was not anything made that was made' (this was the basic assumption of all metascience until, as a complement to his falsifiability interpretation, Anaxagoras II (Popper, 1963) put forward a doctrine of $νου\psi$).

These three types of metascience form their compounds in different ways, however. Instrumentalism forges a tool, more versatile than a crystal ball, by which one can tell of an observation to come. Conventionalism weaves a pattern which might conceivably reflect the actual pattern of things in the world; but since we cannot find out whether this is so, we treat the pattern as a convenience — a convenience for predicting a new piece of the world. Induction builds an edifice out of observations, block by block; the edifice consists of nothing but the building blocks of observations, but the aim is not primarily to move on to new buildings, but to obtain the edifice which constitutes 'knowledge' of the component blocks.

Here, then, are two contrasting philosophies of science — instrumentalism and conventionalism on the one hand, which are purely predictive, and induction on the other, which tries to gain knowledge of the real world. I will now characterise them more closely.

In the first class of interpretations, we have a theory that somehow enables us to predict observations, connect them with one another, pass from one to another. This amounts to a way of handling or *manipulating* observations. The methods are different in detail — instrumentalism for which science is theory-instrumental, and conventionalism for which science is a pattern-theory — but they share this general characteristic. So I would summarise this approach as *observation-manipulation*; alternatively I would call it, to use a term from social anthropology, a *functional* approach to science (insofar as one is not concerned with a societal custom in itself but only with its role in a society). On this approach to scientific theory, one is concerned only with its role in science — not what it is but what it does.

Popper's excellent term 'instrumentalism' covers one of these, but we need a wider term to cover both it and the other two approaches (conventionalism and induction) considered above: I shall call them all 'functionalist'.

Now turn again to the other class. Inductivists do not try merely to manipulate observations; they want observations to grow into something, to grow into a truth wider than themselves; in other words, they want to make the observations produce theories. The contrast is between *observation-manipulation* and *observational construction*. Induction, as well as being functionalist, is concerned also with the *structure* of scientific theory. I would call this type of interpretation of science structuralist; it could also be called ontological.

Thus, all three interpretations are concerned with the problem of getting knowledge out of observations conceived as data (though 'knowledge' is conceived of differently, in one case as 'knowing how to do' something, in another as 'weaving without dropping stitches', in the third as 'construction work'). They find a gap between data and theories: and they provide alternative ways of coping with the gap.[1] Inductivists down to the nineteenth century, to Mill, and up to the present have tried to fill this gap, that is, to pass from data to theories, by some principle of uniformity of nature, later decked out with sophisticated devices involving probability; but they failed to solve the problem of induction, despite their highly ingenious attempts. The classical theory of induction conceived of it as an algorithm:[2] if one followed Bacon's or Mill's rules, making appropriate observations and turning the handle of the induction machine, out would come a generalisation which would be a scientific truth. But the hope of making induction a justified algorithm had to be abandoned. All logicians knew that in itself, unassisted by some rabbit out of the hat, induction was invalid, though they mostly hoped that somehow they would get around the difficulties, that is, find a rabbit in the empty hat. Faced with such a situation, what would an enterprising thinker, coming to the scene afresh and perhaps without the traditional background of logic, try to do? He would try to fill the data-theory gap in some altogether new way. Thus I am suggesting that conventionalists and others were reacting to the same problem of the gap. Why? My evidence is this: if a 'functionalist' regards a theory as being just a compendium or pattern of observations, as stating nothing more than observations, as a sophisticated version of a class of observations, then what he is doing is in effect trying to close the gap by means of an interpretation denying that the gap exists. This may seem to be a slight exaggeration with conventionalism, because with it a scientific theory is apparently broader than the data it encompasses, for it is allowed generality; but when one tries to press conventionalism about what it commits itself to, it makes no claim to being anything beyond its data; for, if 'falsified', it is not a 'false' theory with a wide claim; it is a convention as wide only as the data that satisfy it. Thus, all versions of functionalism equate a theory with some total of observations; and in this way the gap between data and theory ceases to exist. With the instrumentalist it is no exaggeration, for he denies being concerned with reality, so there is no gap between data and theory. Thus, when the gap could not be closed by induction, alternative attempts were made to deny it.

In other words, faced with a failure of the structuralist approach, one functionalist attempt would be to interpret a theory as a set of observations, thus closing the gap by denying that it exists. For although conventionalism does ostensibly appear to make a wider claim than can be justified simply from the data a theory covers, on closer examination it would seem that a theory in the sense of a convention is as wide only as the data that satisfy it. Thus these two groups, the functionalist and the structuralist, are reacting to the same problem — observation conceived as the foundation-stone of knowledge.

In short, the present metameta-conjecture is that both approaches, the functionalist and the structuralist, are responses to a gap between data (or observations) and theory. Where there is a gap which there is no clear way of bridging, there are two obvious devices for handling it: by treating it as non-existent and by jumping it. The functionalist, I suggest, has taken the former way out; the structuralist the latter.

Both take for granted that the process is 'data→theory', whether it is a heuristic gap to be jumped to make a scientific discovery or a logical gap requiring logical justification (cf. Lakatos, 1968). (Instead of trying to prove the axiom of parallels, one tries a new start by denying the axiom and introducing a different approach to parallels.) Instead of jumping or justifying the gap from data to theory, one might try a new start by introducing a new view of the gap (though not necessarily by simply reversing the direction to 'theory→data'.)

I want to emphasise that the two contemporary interpretations of science of this century advocated by scientists, that is, instrumentalism and conventionalism, are modes of observation-manipulation which deal only with the functional aspects of science, and disregard the structural or ontological. All purely functional approaches equate a theory with a sum of observations. Those with a more philosophical background have favoured induction; those whose primary work is in the natural sciences have favoured a theory of observation-manipulation or functionalism. Thus instrumentalism and conventionalism are not alternatives to the observationalism of induction. The fundamental assumption that sets the problem for both approaches, or all three approaches, is the givenness of observations; observations are simply there, as the empiricist tradition has it, waiting to be recorded by that classical camera obscura, the human mind. In other words, they share a philosophy of observationalism.

The falsifiability interpretation, introduced in 1934 by Popper, does not involve observationalism. If, therefore, it can be shown that observationalism is a fatal, irremediable defect in all the standard approaches, then Popper's approach, whether or not it needs modification, offers a way out. Until quite recently it has been an outsider. Within the 1960s it had encroached on established scientific circles (for example, eminent figures such as Bondi in physics and Medawar in biology) and has appealed to many of the younger generation of American philosophers. It is now well established as one of the standard interpretations. It is hypothetico-deductive, like instrumentalism and conventionalism, and in this respect favoured by physicists; it is a theory of testability; it accords with the practice of science. All this is obvious. What is open to consideration is its relation to conventionalism. For, although conventionalism has been portrayed above as faulty, it has, as also mentioned, strong features, and the difference between it and falsifiability may appear small.

In this connection, there are for both problems still to be solved concerning truth and reality; but, fundamentally, it would seem that the choice between falsifiability and conventionalism or the further development of either hinges on the part played in the theoretical structure by observationalism.

It is not my aim here to criticise observationalism. This will be taken up later. My aim is to show that all three of the standard philosophies of science that have been current in this century fail to solve the problem of the data-theory gap and that observationalism is a fundamental assumption of them all.

Notes

1 The way the situation is conceived here echoes Agassi's exegesis. And so does my main dichotomy between conventionalism and induction or between functionalism and structuralism: Agassi's main contrast was between conventionalism and induction. And my stress below on observationalism parallels his on sensationalism. Perhaps our respective conclusions and aims are somewhat different.

2 I owe the notion of an algorithm in a very wide sense to Agassi (1966) though he introduced it in a different setting.

References

Agassi, Joseph (1966), 'Sensationalism', *Mind*, 75, pp.1–24.
Lakatos, Imre (1968), 'Changes in the Problem of Inductive Logic', *The Problem of Inductive Logic*, North-Holland, Amsterdam.
Popper, K.R. (1963), *Conjectures and Refutations*, Basic Books, London and New York.

6 The demarcation problem

Science excludes mathematics and is restricted to empirical knowledge of the world of experience; the overall approach is that only empirical science provides knowledge of the world and affords some hope of authentication. But other enquiries lead to endless disputes for which there is no settled method for arriving at an objective decision in favour of one view or another. Now theoretical science and particularly natural science and still more particularly physics, provides a model for understanding the structure of empirical knowledge. Since theoretical empirical science could produce objective knowledge while other modes of enquiry could not, there is a problem of distinguishing the two themes of science and non-science. This is the problem brought to light by Popper (1934) and he called it the demarcation problem.[1]

Popper's contribution to the solution of this problem was entirely new.[2] The problem itself was to a large extent new but not wholly so. There are two qualifications that should be made about the novelty of the problem. Naturally all sorts of writers had characterised science down the ages and, so far as they tried to characterise it, they in effect demarcated it from other studies. They dubbed it a mode of enquiry carried on by the experimental method, or they characterised it as a mathematical enquiry, etc. None of this is particularly important because such comments were not made with the demarcation problem in view, but were offered merely out of deference to the historical tradition that one is supposed to define one's subject matter, and in addition to hint at an explanation of why the new science was so successful. And this last of course contains the demarcation problem implicitly, but it was never explicitly formulated or indeed focused upon.

There is, however, a much older form of the demarcation problem which was much more nearly explicit. This emerges from Bacon's treatment of the subject in his philosophy of science, for he was overtly concerned to oppose the obscure and undecidable speculations of the middle ages and replace them by modern scientific methods of enquiry. Thus Bacon was very nearly concerned with the demarcation problem: for he could not concern himself with his own main problem without implying, as a corollary, that he was characterising the difference between science and non-science. Nevertheless the demarcation problem was not even then fully explicit, for Bacon's aim was focused slightly differently. It was to characterise scientific method in itself in order to teach how it should be carried out. His aim in other words was to achieve knowledge, to lay down the strategy and tactics by which knowledge could be attained, rather than to tell us the difference between scientific knowledge and other modes of enquiry. Likewise Hume, as Popper has pointed out, made a demarcation, though between sophistry and fact; his emphasis would seem to have lain not on solving the demarcation problem as such so much as on reliability of knowledge so as not to become a victim of sophistry. Popper himself attributes the demarcation problem mainly to Kant, but this hardly accords with Kant's basic demarcation, which was between knowledge of phenomena and the absence of knowledge about noumena, where scientific and philosophical knowledge both lay on the *same side* of the demarcation line. Thus in its full articulation the demarcation problem is properly to be regarded as Popper's problem.

Why was it so important to Popper to effect a demarcation between science and non-science? He grew up in the period of the Einsteinian revolution, which overthrew the greatest scientific theory there had ever been. And it did so on rational grounds. But he also grew up in a time of political and social turmoil. Politically there was Marxism; and it did not for Popper carry the same scientific weightiness as he found in Newton (even though Newton was wrong) and Einstein. Likewise in social thought he found the work of Freud and more especially of Adler wanting in scientific criticism of their own theories.

Popper ran the difference to earth in his theory of testability. Newton's (right or wrong) and Einstein's theories were empirically testable; Marx's and the psychodynamic theories *he considered* were not.

In this book I do not wish to discuss this matter, since it belongs to the philosophy of the social sciences. Nonetheless I will add two points. Popper's relation to Marx and Freud is important because it helped to set him on the road to his metascience — so invaluable for the natural sciences. Properly applied it is, I believe, equally or even more valuable for the social sciences. But curiously Popper misapplied it. He rejected Marx, though some of Marx's conclusions have turned out false, and therefore are at least, on Popper's criterion, scientific. And he rejected Freud and Adler without waiting to see what testing might be attempted — it is as if he knew from the look of them that they were untestable. Be that as it may, Marx and Adler (and Freud) were for Popper untestable and were therefore examples of pseudo-science.

While Popper's problem arose primarily because he saw a need to distinguish science from non-science, it also arose for him in a secondary way because he found himself in opposition to an implied answer to it provided by the logical positivism of the Vienna circle. Like Bacon, the logical positivists were not very far off the demarcation problem, but the focus was different. While Bacon was concerned to point the way to the strategy for obtaining knowledge, the logical positivists, concerned with the method of verifiability, repudiated metaphysical knowledge, which they held to be meaningless. They were concerned with the philosophy of science not so much for its own sake as to provide themselves with a method for eliminating meaningless assertions. Their way of doing this was by means of the principle of verifiability, according to which a theory, in order to be meaningful, would have to have deductive consequences which would be amenable to empirical verification. The stress here is on verification. If Popper's stress on the demarcation problem had a secondary source, it also had a secondary consequence, for his solution provides a general strategy for developing scientific knowledge.

Popper found himself opposed to Bacon because the inductive method was invalid and found himself opposed to the logical positivists because the method of verifiability was logically untenable. A detailed summary of numerous defects of induction was given in the appendix to Chapter 2, but a brief comment on verifiability may be made here.

As a criterion for empirical science the principle of verifiability does not suffice. Suppose you entertain the hypothesis that a certain sort of drug will prevent seasickness; you carry out a logical deduction and infer that if you take this drug you will not be seasick. To verify this take a trip in a boat after swallowing the drug and spend the day at sea; now it turns out that it is a fine summer day without a ripple and you return home feeling perfectly well. Since you are well you have obtained a verification of the hypothesis. The absurdity of the result suffices to show that the verification involved has no bearing on the truth or falsity of the hypothesis. It points the way, however, to an improvement that can be introduced. You will immediately reflect that what you should have done was to make sure of taking the boat on a rough sea, and then, if you came home feeling well, you would have grounds for confidence in the hypothesis. When we consider this situation, however, we notice about it that the second experiment is conducted in such a way that the outcome could have rendered the hypothesis false, that is, shown it to be false if indeed the drug had been no good, that is, if the hypothesis had been false. Thus the shortcoming of the verifiability approach is that, by itself, it is wholly useless in conditions where the experiment or tests or situations could not possibly go against the hypothesis. In other words, it is serviceable only if the hypothesis was investigated in circumstances which could have refuted or falsified it. Then having reached this point we notice that falsification suffices as a characterisation; and verification adds nothing to it at all. Hence we have Popper's notion that the characteristic of science is the possibility of falsification or refutation. In other words a theory to be scientific must be in principle falsifiable or refutable.

Popper put this forward as a theory of the nature of all science, so far as universal explanations go. He extended it also to cover particular explanations of a historical kind, but that is not our concern now. In sum, a universal statement is scientific if and only if it is empirically refutable or falsifiable in principle.

There are some subsidiary qualifications to be noted. It may very well be that a given theory is unfalsifiable in practice because no one has been able to think of a way of carrying out a possible falsification. This does not show that the theory is unscientific; it shows only that we do not yet know. It is easy enough to slide from unfalsifiability in practice to the intuitive assumption that a theory is unfalsifiable in principle. Popper does occasionally do this, though rarely, as for instance when he writes off psychoanalysis and certain features of Marxism as unfalsifiable before any serious attempt has been made to enquire whether a falsifiability procedure might be found. There, are, however, some statements of a very peculiar character which can be seen by inspection to be unfalsifiable because they have built into them a clause that for ever prevents falsification from taking place. These will not occupy us until later.

We have to elaborate Popper's criterion a step further. A universal statement is scientific if it is in principle falsifiable by an empirical observation. The raison d'être of falsifiability is testing by means of observations which are intersubjectively accessible, that is to say, accessible not just to one person but to anyone, at least anyone with the appropriate competence to make the observation. Popper himself, it is true, expresses the matter slightly differently by dividing the test procedure into two stages. Firstly he holds that a general theory is falsifiable if from it can be deduced an empirical generalisation. The above satisfies the same requirements. Now, secondly, an empirical generalisation is falsifiable in the same way by a possible observation. For most purposes, however, in the philosophy of science there is no point in dividing the process into these stages, and a logical consequence of Popper's procedure is the collapsed version given above: it will serve as a characterisation of a universal scientific statement or theory. The most important distinction to be brought in at this point is the non-absoluteness of falsifiability.

Popper has made it very clear that verification cannot be absolute: although it is possible to derive from a theory consequences infinite in number, yet we never come to the end of these; thus it is impossible for the hypothesis to be confirmed absolutely no matter how many verifying or confirming instances are found. By contrast there is asymmetry between verification and falsification here, for while no number of verifying instances will show a theory to be true, *one falsifying instance can show a theory to be false*. In other words, one concrete example will knock out a theory.

This is a purely logical point, and the reason for emphasising it is solely to show the difference between falsification and verification, that is, that they have a different logical structure. To say that one falsifying observation would knock out a theory holds only in a certain setting, namely, a setting in which none of the other circumstances involved is in question. But a theory may be accompanied by other premisses and falsity may be attributed to them rather than to the theory. Hence we can never be certain that we have falsified a theory.

The non-absoluteness of falsifiability is more conveniently discussed in the next chapter in connexion with Kuhn, who raised the question in an acute form.

Appendix: Popper's Metascience of Explanation

It may be worth giving a rudimentary axiomatisation of Popper's metascience of explanation. Popper put forward his view of hypotheses, deductions, refutability, independent lines of testing, and holds that he is giving an account of 'explanation'; he also stresses the flaw in ad hoc hypotheses. One could, however, put the system together like this:

> Axiom 1: A hypothesis is not an explanation if it is ad hoc.

For an 'explanation' that was fully ad hoc, the 'explanatory' hypothesis would be identical with what was to be explained.

> Axiom 2: A hypothesis is not an explanation if it has less information content than (or an equal amount to) the explicandum.
>
> Theorem 1: To be an explanation, a hypothesis must have more information content than the explicandum.
>
> Definition 1: An explanation has explanatory surplus if and only if it has more information content than the explicandum.
>
> Theorem 2: To be an explanation, a hypothesis must have explanatory surplus.
>
> Theorem 3: Consequences other than the explicandum are deducible from an explanation.

Popper has himself shown very clearly that consequences which can confirm only, without the possibility of refutation, do not support a hypothesis; and this could easily be axiomatised. What comes out less easily perhaps is that a confirmatory consequence (even if it could have refuted the hypothesis) gives no support to that hypothesis unless in doing so it would refute an existing rival hypothesis. (Otherwise expressed, a consequence of this sort gives no support if it is one you would expect anyway even without the hypothesis under test; it is rather like saying that a man cannot claim to be the father of a certain child if the child would have been there even if the man had not existed.) This point is sometimes overlooked by those who make a serious attempt to apply Popper's theory.

This requirement is, so to speak, the complementary opposite of that about explanatory surplus: if an explanation has to be distended, a consequence or evidence must be reduced or sieved; that is to say, nothing can be allowed through as evidence if there is already an explanation for it.

Definition 2: Evidence is said to be 'independently expected' if and only if it is the consequence of another known accepted hypothesis.

Theorem 4: A hypothesis with a consequence that is 'independently expected evidence' receives no support from it.

Definition 3: 'Independently expected evidence' is said to be 'irrelevant' (or to have 'no corroborative power').

However this may be, and whatever the outcome of the Popper-Carnap controversy, it is plain that metascience has more or less 'arrived'.

Notes

1 Wallis (1984) has given a closely reasoned argument to show that no proposed criterion successfully demarcates science from pseudo-science. His main point against Popper is: to be a falsifier an observation presupposes a subjunctive conditional, which is that if the test were repeated the observation would do as it had done before — a generalisation that could not be tested (and would in effect be equivalent to an induction).

I think Popper would not agree, but hold that for pragmatic purposes it would be idiotic to use a falsified hypothesis, and though a survived hypothesis cannot be guaranteed, it is a more sensible tool. And so far as practical use — and also theoretical explanation — goes, Popper always holds that a survivor hypothesis may always fail in the future.

2 My former student, Dr Tom Bowers (c.1975), has rightly pointed out to me that the demarcation problem was, according to Popper, known to Hume (it was also virtually known to Bacon), and also that it was the central problem in Kant's theory of knowledge. While Popper's problem mainly concerns the difference between scientific and non-scientific knowledge, that is, with characterising scientific procedure, it also has a bearing on a topic pursued below in Chapter 17, that is, on the growth (or otherwise) of scientific discovery.

References

Bowers, Tom (c.1975), Personal communication.
Poincaré, Henri (1952), *Science and Hypothesis*, Dover, New York.
Popper, K.R. (1934), *Logik der Forschung: zur Erkenntnistheorie der modernen Naturwissenschaft*, Springen, Wien (imprint, 1935).
Popper, K.R. (1959), *The Logic of Scientific Discovery*, Basic Books, London and New York.
Wallis, Roy (1984), 'Science and Pseudo-Science', paper to Hist. and Philos. Sc. symposium, Royal Irish Academy, Dublin, 18 April.

7 Theories not absolutely falsifiable

When Popper put forward his thesis that to be scientific a statement must be falsifiable, he made it very clear that no theory could be falsified absolutely. This evident fact has often been ignored or treated as inconsistent with his overall thesis. It is important to get his thesis straight, and this can benefit from certain additional points he did not bring to bear. The matter arises afresh and in a convenient way because of the contribution made by Kuhn's (1962) extraordinarily able and interesting work *The Structure of Scientific Revolutions*. The outstanding feature of Kuhn's book is that it is written by a scientist who looks through the eyes of a scientist, knows what scientific work feels like, both in its dramatic discoveries and its day-to-day work — a point that is sometimes overlooked by metascientific critics. Reflection on Kuhn's work will enable us to arrive at what could be Popper's metascience of the history of science. The problems to be discussed are scientific crisis and the relation between logical corroboration of a theory and its sociological acceptance. Although at first sight it seems that Kuhn is doing something quite different from Popper, there is, despite some disparity of views, an extraordinary likeness.

The role of a counter-example for Popper and Kuhn

Kuhn has disagreed with what he took to be Popper's view of the effect of a counter-example, and makes, I think, one of his very few errors. He remarks, quite rightly, that if something goes wrong with a test or some counter-example turns up, this is not immediately regarded by a scientist as a counter-example to a fundamental theory (Kuhn, 1962, esp. pp.77, 81). But Kuhn supposes that, according to Popper, the counter-example is a counter-example to the theory and refutes it. Kuhn disagrees with the supposedly Popperian contention (1959, p.145) and says, rightly, that the scientist would, of course, be right. For the scientist a whole crop of questions would arise. In the case of an experimental test involving mathematics, the scientist would doubtless first check the mathematics, then enquire whether some accidental disturbance had spoilt the experiment, then check whether his instruments had been working properly, and then look into the possibility of bungling the experimental work; he might even begin to wonder whether there was some subsidiary hypothesis about the use of the instruments that was wrong, and he might go the length of considering whether the fundamental theory was correctly articulated for the purpose in hand. Many steps would have to be taken, which meta-scientists scarcely ever mention (if indeed they ever noticed them), before coming to those that interest them. So, Kuhn is right that a counter-example is not a refutation of a theory.[1] But Popper knows this too. Kuhn, somehow, perhaps understandably, thought that Popper would deny it.[2]

Let us now dwell on a relevant point in Popper's (1959, pp.40f) theory of refutability, which is quite clearly stated in his opening work of 1934, and which has not been modified since on the point at issue. The theory of refutability in its intuitive outline is very simple. As we saw in the previous chapter, a theory to be accredited as scientific has to be refutable, and the conditions for refutability have to be specified. This means that, if a consequence of a theory turns out to be wrong, then the theory is falsified. This is the rough intuitive way one can put it to begin with. Now this is misleading, for it might suggest that the falsification was absolute; but Popper (1959, pp.42, 47, Ch.5) holds it is not; and he amplifies his theory in the body of his work. One has the theory as a major premiss. But other premisses are held as well — perhaps several of them. There must be at least one other, which will state initial conditions; for it is impossible to derive an empirical consequence (or make a prediction) from a theory without bringing in initial conditions, or some form of observation-statement. Thus you have several premisses, at least two, when you make a test. When you get a prediction that is wrong, this tells you that, if the inference is valid, there is falsity to be unearthed and pinpointed somewhere among the premisses. But it does not tell you where. Popper knows this perfectly well. He has emphasised it many times, it is in his first book, and he never overlooked the point. You do not know that the theory is falsified. You do know only that one of the premisses is false. That premiss *might* be the theory, but it might be the initial conditions. And this is as far as you get with the falsification.[3]

Refutability-theory compatible with 'normal science'

We may now enquire how far this development of Popper squares with Kuhn's theory of 'normal' science and then with his theory of crisis.

The question arises for the working scientist, however, though not for the metascientist, whether the error is to be pinpointed on one of the other features mentioned above. Was there a mistake in the logical deduction (Kuhn, 1962, p.81)? Or a chance effect that altered the initial conditions? Or some failure of instruments, bearing on one of the minor premisses? Or some bungling which would do the same? Or some inaccurate articulation of the theory? Or what? The position may be put more sharply by classifying the premisses or other features of a scientific inference as follows: (i) the general premisses constituting the fundamental theory; (ii) the particular observational premisses constituting the initial conditions; (iii) subsidiary premisses/assumptions to do with miscellaneous conditions concerning chance effects, failure of instruments, theory of instruments, bungling, articulation of theory, validity of deductions. It seems to me good horse-sense, which anyone doing any kind of practical or theoretical work would always do, on the grounds of economy of time, economy of labour, and perhaps for other reasons, that the scientist — though not the metascientist — will first concern himself with the possibilities of loopholes in the subsidiary premisses/assumptions, in case something will turn up that will save the prediction, save the theory. In other words, he will try to do puzzle-solving.

The only reason the metascientist is not at one with this is not because of disagreement but lack of interest: he assumes that such steps have been tried (assuming he knows about them), that is, that the subsidiary premisses/assumptions have been checked, and he becomes interested only when these steps have failed to work out the puzzle. What happens then? Supposing you cannot find anything wrong with the test, do you throw out the fundamental theory? The answer is still 'No', both for the scientist and the metascientist: the fault may be in the initial conditions. In this situation, I think that, rationally speaking, the only thing to do is to suspend judgement. It is too early to know whether the fundamental theory is wrong, or whether something undedected about the initial conditions is involved. (The latter turned out to be the answer over the perturbations of Uranus, which were explained when the initial conditions were altered by Adams and Leverrier to include Neptune.)

Thus, even at this stage, a false consequence of a theory is not a counter-example to the theory. Even when the subsidiary premisses/assumptions have been gone over, it is a counter-example only to the remaining set of premises/assumptions yielding it. How does the scientist settle the question at this point of whether the fault, or falsity, lies in the theory or in the initial conditions? Again there is no specific metascientific prescription for this.[4] Falling back once again on the notion of economy of time and effort, when a businessman considers how to handle a difficulty, the sensible thing to do is to work on the initial conditions first, simply because this is much easier than the immensely difficult task of altering the theory.

I propose to call this phase one of 'paradigm-exploitation' (to anticipate a conception of Kuhn's) to cover elaboration, puzzle-solving, theoretical application, etc. I have presented it as compatible with Popper's refutability-theory of metascience, without interpreting the phase as one of testing underlying theory. Thus Popper's metascience is entirely consonant with Kuhn's phase of 'normal' science.

From anomaly to crisis

After serious attempts at puzzle-solving have failed, we have what Kuhn calls an 'anomaly'. Suppose an anomaly occurs a handful of times. Then you have a situation that Kuhn describes as a 'crisis' (Kuhn, 1962, p.82). At this juncture, scientists do not know where to turn next.

It seems to me that this position is exactly what is required according to Popper's metascientific theory and Kuhn's theory of history. At this point one can get aid from ordinary simple considerations about what is most likely. Given a number of situations in which predictions have gone wrong, say a handful, then it becomes highly unlikely that in every single one of those there is some mistake about the initial conditions (or even that there is some unlocated mistake in the subsidiary premises/assumptions); this becomes more and more unlikely with each particular anomaly that occurs. So, with a handful of these, the probability that they can *all* be explained in this way is extremely small. It is at that point that it becomes reasonable for scientists to consider that perhaps, after all, the fault lies not with the initial conditions nor with other premises but with the fundamental theory. It is at this point that the false consequences or counter-examples can be regarded as a refutation specifically of the theory (though the refutation is not absolute).

The position is the same on Popper's metascience; for, on his view, falsification is not absolute, though often taken to be so. It has sometimes seemed to philosophers that, when Popper stressed the impossibility of obtaining verification of a hypothesis no matter how many confirmations were found, whereas one disconfirmation would falsify it, this implied the absoluteness of falsification. But this is so, for Popper, only if the initial conditions are taken for granted and are not in question. I have stressed that this is a logical point. His theory puts falsification in a totally different position from verification: verification is impossible even if the initial conditions are granted; falsification is possible (and would be absolute) if the statement of initial conditions is true. Thus the concept of falsifiability has applicability, even though it cannot be applied with absolute certainty. In the real situation, the initial conditions may be questioned and then the falsification of a hypothesis is not absolute (Popper, 1959, pp.42, 47, Ch.5).

It must be emphasised that there are two distinct problems: that of demarcation, in relation to which verification cannot be completed while falsification is absolute; and that of the non-absoluteness of a refutation because of the possibility of questioning the initial conditions etc. This distinction was clear in Popper's original 1934 work, but three decades later he explicitly contrasted them somewhat as I have done (Popper, 1963).

On Popper's theory, then, one can accept Kuhn's thesis that there is a period of science consisting of elaborations, exploitations, applications, and things of that sort, in which the fundamental aim is not the testing of a theory. When there are sufficient failures, that is, there are enough anomalies to create a crisis, then the anomalies *become* tests of the theory and of the fundamental premiss of the theory rather than of the other premisses.

I think I have at one and the same time managed to outline both Kuhn's theory and Popper's, because the apparent difference between their views largely evaporates in the way I have put the position. This result hinges on the fact, which Popper himself made explicit even in his first work, that falsification is not absolute, or that a false consequence is a counter-example only to a set of premisses and not necessarily to the premiss expressing the theory (Lakatos, 1967).

Kuhn's sociology of acceptance of new theories

I should first try to bring out such differences in their views as Kuhn and Popper might suppose were there. For Kuhn there is an extralogical factor in the situation: the scientist simply does not wish to give up the theory when an anomaly first arises. But, when there are several anomalies, there is a crisis, and maybe a new theory is produced. The new theory cannot be adopted on any kind of logical grounds, for it creates too many new problems and there is no knowing whether the newer problems will be too difficult for it, and it is adopted at least partly because it overcomes the immediate crisis (Kuhn, 1962, pp.155–7). Popper's theory, by contrast, looks as if it depended solely on logical considerations. Thus Kuhn takes a view of science that appears to be slightly less rationalistic than Popper's, in the sense (he does not speak of 'rationality') that he holds there is no specific logical procedure, nothing resembling a demonstration, that compels rejection or acceptance of a theory. Before discussing this, let us consider a cognate point.

Kuhn (1962, pp.146—50) gives the impression first of all that scientists who develop a new theory cannot be on speaking terms with scientists of the first theory — to be specific, that Einsteinian scientists could not really understand Newtonian physics, and the other way around. (This impression does, I think, more or less reflect Kuhn's view, but it should be borne in mind that he allows some role to logical considerations.) It certainly is true in practice that there is often great difficulty in communication between scientists who follow a new theory and those who follow an old; but certainly there are times when they can sift the matter out and understand one another. Nonetheless he does give a certain amount of powder and shot to those who criticise him for giving an irrationalistic account of scientific development. I think it is possible that, in his skilful depiction of the entire change, not only of theory, but of constituent concepts (e.g. mass) from Newtonian mechanics to relativity (Kuhn, 1962, pp.100—1, 148ff.), he may have gone beyond what was needed for this purpose and suggested the impossibility of communication in principle.

Kuhn (1962, Ch.12) adds to the somewhat irrational flavour by suggesting that when a new theory is produced there is a bandwagon effect (though this expression somewhat exaggerates his sociological tendency). Somehow or other, the new theory becomes accepted, and the old one becomes rejected, as a result of a not fully rational change of attitude. Now there are, of course, all sorts of social reasons why this is so. A new generation of scientists grows up, and the old ones die off or retire, and we all know that in middle age we get hardening of the arteries and find it more difficult to understand new ideas. This is a commonplace. Such facts give a sociological basis for the idea that scientific theories do not get refuted; like old soldiers, they only fade away. This is the sort of impression Kuhn conveys.[5] However, the question is whether such sociological pressures are a fundamental or merely an auxiliary effect that takes place on the surface, or is there also a rational factor involved, and is it decisive? I think, myself, that there is a different process at work, which I now wish to describe. And I shall at the same time try to show how Kuhn tries to include — I think unsuccessfully — logical considerations within his sociological theory of rejection and acceptance of paradigms.

Although Kuhn allows a role for refutation, he is somewhat ambiguous about it and the end-result gives an impression that he wishes to play it down. Certainly Feyerabend and Lakatos discount it. And indeed it suits part of the temper of the times to do so. In view of this, it is worth remarking on the extraordinary nature of this tendency.

What is discussion, argument, and inference if refutation in no sense can occur? Moreover, when for Lakatos there is a shift in the course of a research programme from one hypothesis to the next one which is already lined up in case of need, *why the shift unless the earlier one is in some sense refuted?*[6]

Alternative balancing of logical and sociological determinants

When a new theory is propounded, there is, no doubt, a great shock to the middle-aged, who are perhaps emotionally tied to previous theories; and there may be a great deal of difficulty in understanding the new theory (some people may be able to remember the enormous difficulty people had in understanding Einstein in the first quarter of the century). Nonetheless, in the end the small number of top-grade scientists in a certain line will recognise that the evidence favours the new theory and will accept it on rational grounds.

Kuhn knows this just as well as I do, and he is overt about it. Why, therefore, does he give a sociological account, when he recognises that there are rational factors to do with evidence? What he says is that evidence plays a part. The question is, *what part*; for it must play a decisive part, or the role it plays is not significant at all so far as a specific logical procedure for rejection or acceptance of a theory is concerned. Kuhn's answer is a rather curious one. Let us consider some key passages.

> ... to say ... that paradigm change cannot be justified by proof, is not to say that no arguments are relevant ... (p.151)
>
> All the arguments for a new paradigm discussed so far have been based upon the competitors' comparative ability to solve problems. To scientists these arguments are primarily the most significant and persuasive ... But, for reasons to which we shall shortly revert, they are neither individually nor collectively compelling. (p.153)
>
> ... if a new candidate for paradigm had to be judged from the start by hard-headed people who examined only relative problem-solving ability, the sciences would experience very few major revolutions. (p.156)
>
> Even today Einstein's general theory attracts men principally on aesthetical grounds ... (p.157)
>
> If authority alone, and particularly if non-professional authority, were the arbiter of paradigm debates, the outcome of those debates might still be revolution, but it would not be *scientific* revolution. The very existence of science depends upon vesting the power to choose between paradigms in the members of a special kind of community. (p.166)

> . . . scientists will be reluctant to embrace it [a new candidate for paradigm] unless convinced that two all-important conditions are being met. First, the new candidate must seem to resolve some outstanding and generally recognized problem that can be met in no other way. Second, the new paradigm must promise to preserve a relatively large part of the concrete problem-solving ability that was accrued to science through its predecessors. (p.168)

These passages display a shift of emphasis from the influence of argument to its lack of adequacy, back to a basis in what is scientific, thence to a community-judgement, and back to scientific efficacy. This last seems to be a wholly rational justification, that is, it seems to place the whole emphasis on evidence: if a new paradigm solves problems an old one does not, the old one is refuted. But the role of sociological factors does not allow this to be decisive. How then are we to resolve the apparent ambiguity of Kuhn's position? There is one more passage, which supplies the key.

> Because scientists are reasonable men, one or another argument will ultimately persuade many of them. *But there is no single argument that can or should persuade them all.* (p.157, my italics)

This is susceptible to the interpretation either that the pieces of evidence that appeal to one scientist are not necessarily the same as those that appeal to another, or that the same pieces of evidence are weighed differently. Now, if you will just dwell on this for a moment, it is much more curious than it looks. Supposing *you* are possessed of, say, three pieces of evidence that convince *you*. And suppose *I* am convinced by three different factors, or at any rate suppose not all of these are the same. Then there is agreement among the various scientists on the grounds of evidence, but there is nothing that you could call inter-subjective testing. And the same holds if very different weights are attached to factors they agree upon. So, although they may be rational men and thinking in terms of evidence for the theory, since there is lack of agreement about the evidence, specific logical factors though present play no decisive role. You have a situation that I would describe as being like a syndrome of a disease, where the set of symptoms varies from patient to patient, or, if the same, varies as regards intensity, and is the same in several patients only in rare cases. The doctor makes a diagnosis on the basis of a syndrome which is not the same as the one the textbook describes as typical, or differs from it markedly in intensity. This is commonplace even in an epidemic of influenza. Of course, you hope that there is one characteristic feature which will reveal the disorder quite decisively, but this is very often not the case. Thus there is a question of skill in comprehending a syndrome which does not follow the usual description in the textbook. The doctor nonetheless does have an independent test, if he makes a diagnosis on the basis of a varying syndrome. He can test it by sending certain specimens to a laboratory for further examination.

But, in our situation, consisting of a varying syndrome of evidence, there is no independent test, no constant intersubjective observation or evidence. So despite Kuhn's holding that evidence or some specific logical factor is at work, or plays a role, it plays no *decisive* role.

Therefore, it seems to me that fundamentally Kuhn comes down on one side of a knife edge, in favour of sociological explanations of scientific change.[7] Popper's metascience would come down on the other side of the knife edge, though there would, of course, be no reason to deny that sociological factors entered in as auxiliary influences. On Popper's metascience, it would have to be the same factors in comparable degree — the same evidence — that appeal to all scientists who are really competent to judge. And when they have let it be known that they accept the theory, then I think it is at that point that the bandwagon effect may take place. It convinces people on the fringe that they need not worry about the controversy any longer. Supposing, for instance, that you are working in low temperature physics, but are not a specialist, say, in a theory of motion, you may not be too interested in going into the question of whether the tests in relativity are genuine, convincing, and decisive. But, if the recognised authorities on it are satisfied, then you tend to be, too. This is reasonably respectable and, in calling it a bandwagon effect, I do not want to cast too much of an aspersion on it (though there can of course be less noble bandwagon effects as well). Thus, so long as you have a specific logical procedure as a basis, e.g. uniform evidence accepted by the top-grade people in the line, it does not affect the rationality or the status of the theory if a lot of rank and file people, or top-grade people in another line, accept the theory mainly on sociological grounds.

Conclusion

With these three sections I conclude my comparison of Kuhn and Popper with respect to exploitation of paradigm, crisis, and corroboration *versus* sociological acceptance. On the first two I have tried to show that they are fundamentally at one. On the first, to do with 'normal science' or paradigm-exploitation, I have reconstructed Popper in a way that aligns him with Kuhn. On the second, concerning crisis and revolution, I have given a reconstruction that both Popper and Kuhn can fit. On the third there is a knife-edge difference which creates a deep and fundamental cleavage; I have given a reconstruction which Popper can fit, but which, though it absorbs an element from

Kuhn, is incompatible with his sociological theory of scientific knowledge.

One of the main objectives of this chapter was to show clearly in what way a theory is not absolutely falsifiable, while showing that the falsifiability thesis is not thereby rendered useless. This enables us to see that Popper and Kuhn can be brought extremely close together. But it also enables us to see the fundamental gulf between Kuhn and Popper — a gulf, even though it results from what I have called a divergence at a 'knife edge'. For Popper, even when developed as I have done above, science remains objective; for Kuhn, despite his attempt to take rational factors and scientific evidence into account, it is after all subjective.

When a counter-example turns up, Kuhn claims it is not a counter-example to the fundamental theory and claims that scientists never treat it as such. I claim Popper's metascience likewise opposes this. Before such a step is taken, a whole crop of loopholes have to be investigated: checking mathematics, accidental disturbances, instruments in working order, bungling, unsuspected misconceptions about the nature of the instruments, adequate articulation of the fundamental theory. We have to see that this is in line with Popper's theory. He stresses the role of initial conditions; so, for him, a counter-example is a counter-example to an inference containing not only the fundamental theory but also the initial conditions. There are also the other conditions listed, which may be called subsidiary premisses/assumptions. Metascientists are not interested in these; they are interested in the situation only when these subsidiary premisses/assumptions have been investigated. Even then the counter-example may point to an error, not in the fundamental theory but in the initial conditions. Only when the initial conditions have been investigated is it correct on Popper's metascience to consider pinpointing the error on the fundamental theory. But even then this cannot be done with complete certainty, for some other factor may have been overlooked. So long as the fundamental theory is protected, puzzle-solving and paradigm-exploitation continue. When all the minor premisses that can be thought of, including the initial conditions, have been investigated, the counter-example looks like being a refutation of the fundamental theory, but in view of doubt it is not yet taken to be so. It has the status of being an 'anomaly'.

However, when you get a handful of anomalies, it becomes highly improbable that something has been overlooked. At this point the handful of anomalies constitute a 'crisis' and the fundamental theory (empirical content) is taken to be refuted (though not with complete certainty).

It is, I claim, the same with Popper; for falsification in his metascience is not absolute.[8]

Thus Kuhn's theory and (when correctly interpreted) Popper's (when developed) are identical.[9] Nevertheless, in the sociological way in which Kuhn sees himself seeing science, they are poles apart.[10]

Notes

1 For simplicity one may write of disposing of the possibility of mistakes in the subsidiary premisses/assumptions and then concentrating on the theory and initial conditions; but in fact we can never dispose of all the possible subsidiary premisses/assumptions.
2 Feyerabend in his more recent work (*Against Method*, 1975) has, alas, swallowed unquestioningly the travesty of Popper widely circulated by Lakatos, without however using Lakatos' notation. (Kuhn had made the misinterpretation earlier.) Lakatos interpreted a $Popper_0$, $Popper_1$, and $Popper_2$, which had to be given up in turn. Lakatos thus failed to appreciate the distinction between the logical role of refutation and its role when applied to a scientific system. I hope that the above chapter has now made this distinction quite clear, and that Popper was not confusing them, and that therefore Popper did not take a counter-example to be a refutation of a theory.
3 Lakatos has discussed Popper and Kuhn from the point of view of what Popper apparently calls metaphysical research programmes (I have not seen Popper's unpublished work on this). This means that, if a theory is falsified, it is well known in advance what step (and succession of steps) will be taken to make an adjustment; thus, if the theory is that motion is circular, the next variant will be that it is elliptical, the next that the centre of force will be moved, and so on; all this within one programme. Such a procedure is highly important as a distinct part of 'normal' science; its characteristics are that refutation of the programme is not absolute (though after a time it may be found impossible to carry the programme a step further and it is dropped). I would classify this phase of science as the childhood of a paradigm. It belongs more to a period just before 'normal' science or just after revolution rather than to revolution itself. Kuhn discussed pre-paradigm science, though not this aspect of it; but he would surely be very ready to make room for it. I do not think it calls for separate discussion in the present context, because my aims concern different issues. I do think, however, that Lakatos has drawn attention to a highly important *phase* in the growth of science. In developing this idea Lakatos would seem to presuppose an expansion of Popper's metascience such as I have developed.
4 Dr Bowers has also argued (as indeed Popper has recognised) that too early a refutation — his example is the early falsification of

Copernicus — might possibly abort the growth of science. For, in conjunction with Aristotelian physics, it would entail that bodies — and the air — would fly off the earth.

5   Hattiangadi (1965) has provided a well argued criticism of Kuhn's sociological theory of scientific knowledge.

6   Kuhn writes (in a personal communication): 'I mean only to be insisting that, where choice of theories is at issue, there is no fully logical argument from agreed empirical evidence that will compel a choice or convict a man of irrationality . . . . Adherents of an old theory can defend it until their deaths by fully rational arguments.' I am grateful to Professor Kuhn for allowing me to quote this passage from a letter (and for other helpful comments and elucidations).

7   Yet they work out closer to one another than meets the eye; it is also worth remarking on the numerous features of Popper's metascience that Kuhn adopts: the need for refutability, the significance of unanticipated novelty, necessity for a new theory to reject an older one, the need to risk being wrong, the role of observation and experiment, the dominance of theory.

8   An attractive discussion, much in line with the above, has been given by Musgrave (1973).

9   Lakatos (1967) has (independently) come to exactly the same conclusion. (Added in December 1971: This note, written over four years ago, refers to views Lakatos put forward in 1967; the present discussion is concerned only with the original clash, where the differences show up most sharply.)

10   Added in 1980: In his reply to the original publication of this paper in *The Philosophy of Karl Popper* (1974, p.1153), apart from one minor qualification, Sir Karl made only one criticism, namely his surprise that I had concluded that Kuhn's thesis and Popper's 'are, when correctly interpreted, "identical"'. This remark of mine referred to their views about the non-absoluteness of refutability, and to the consequential development of Popper's view that I introduced to handle anomalies within Popper's metascience. But it did not mean — and the whole development of my paper shows this — that they were identical over rationality, evidence, testing — my aim was to show the difference. Added in 1983: What did not come over to Popper is that the identity between Kuhn and Popper concerned an *isolated part* of the overall theory, namely the *carefully formulated* understanding of refutability. The identity referred to was not the overall metascientific theory: about this I contended that there was a 'knife-edge' difference (which produces a chasm) in which Kuhn ultimately rejects rational evidence appraisal as the sole criterion of this remains on the sociological side of the divide (however much he allows that rational factors play some — peculiar sociological — part), while Popper does not.

Popper agrees that sociological factors have a role, e.g. in encouraging or discouraging a scientist from accepting the rational evidence, but for him it is the rational evidence that counts.

References

Bowers, Tom (c.1975), Personal communication.
Hattiangadi, Jagdish N. (1965), 'Truth, Acceptance, and Agreement: a Discussion of Prof. Kuhn's Theory of Science', MA Thesis, University of London, Ch.V.
Kuhn, Thomas (1962), *The Structure of Scientific Revolutions*, University of Chicago Press, Chicago.
Lakatos, Imre (1967), 'Demarcation Criteria of Scientific Research Programmes', paper given at the University of Southern California.
Musgrave, A.E. (1973), 'Falsification and its Critics', *Logic, Methodology, and Philosophy of Science*, North-Holland, Amersterdam, Vol.4, pp.393–406.
Popper, K.R. (1959), *The Logic of Scientific Discovery*, Hutchinson, London and New York.
Popper, K.R. (1963), *Conjectures and Refutations*, Basic Books, London and New York.
Sachs, Mendel (1985), 'Quantum Theory and the Schism in Physics', *Philosophy of the Social Sciences, 15*, pp.321–31.

8 The nature of 'normal' science

This topic has arisen partly as a result of an interpretation of it given by Kuhn and partly as a result of new views of the history of science which stem from new theories of the nature of science (Kuhn, 1962; Agassi, 1963). It is, however, a metascientific problem, not a historical one. Though the canvas is the history of science, my discussion has to do with the role of metascience, or the philosophy of science operating in the medium of the history of science.

Let us begin by culling some points from a worthy contribution by Butterfield (1947). For him the history of science — by which he has in mind the history of scientific revolutions — is not just a history of one great name followed by another great name, that is to say, a kind of chart joining up points from 1686, say, to the next great date: narration, or what I would call chronography. Nor is it just a history of success stories. Nor is it a cumulative doctrine that science is the accumulation of discoveries or the accumulation of knowledge of facts. All these things Butterfield decries. More recently many of them — success stories, accumulation — have been subject to heavy fire from Agassi, who has incisively demolished them.[1] Butterfield makes no heavy weather over it. He makes it plain that these are not reasonable approaches to history.

Butterfield's positive thesis can, I think, be put into two words. First, so far from revolutions, being traceable to external factors, the position is that scientists are at some stage floundering with problems in struggling with which they undergo a change in the workings of their minds; they see old things in a new way, and manage to get a *key idea* (this is a phrase he is very fond of); then, with the key the lock is turned, so to speak. Second, when unlocked, the *sluice gates* are opened, by which he means that discoveries then flow along very easily, once you have got the key and unlocked the gates.[2]

Kuhn would, I think, accept Butterfield's two major ideas: the metaphor of keys and the sluice gates, as at least part of the story, though Kuhn goes further;[3] in fact, it will become plain that Butterfield's key is Kuhn's paradigm, and the sluice gates are Kuhn's periods of 'normal' science consisting of puzzle-solving.

Kuhn on 'normal' science

Now Kuhn has a thesis that embraces a number of very different phases, including a period in normal science followed by a period of crisis. Although Kuhn's title concerns scientific revolutions, he gives close attention to what he calls 'normal' science, which is non-revolutionary. And this, I think, is where one metascientific problem arises.

The period of normal science is one which he spreads himself over and gives a very fine description of. This is the sort of thing that you do not readily find in books on the history of science or in books of metascience. I will try to give a bird's eye view of what he has to say about this. In some periods the scientist is concerned with elaborating knowledge he already has, that is, making and getting more accurate estimates of existing knowledge. For example, he may want to estimate the mass of the moon more accurately than was done before. He may want to estimate melting points or densities and things of that sort more accurately, and a lot of careful experimental work may be involved. Such work is both factual and theoretical. It has to do with establishing facts which might be called general facts. One may need some of these facts for certain practical purposes like nautical almanacs; or it may be interesting to predict eclipses with more accuracy than previously predicted. One needs these things for practical purposes and for developments in technology. But one also needs them because they have to do with relations of facts to theories, in particular with the possibility of further testing existing theories. For instance, getting the mass of the moon right is highly important for testing one of the applications/consequences of Newton's theory, namely the motion of the tides: as an application, this provides information about the tides needed for practical purposes; as a consequence, it provides a test (corroboration) of Newton's theory.

That would be one area which could be briefly described as investigating certain kinds of facts *with greater discrimination* than was done before. My main reason for highlighting this point is that it gives a picture of normal science that is usually overlooked by philosophers of science. Another area of normal science concerns crucial predictions or calculations of such things as the acceleration due to gravity at different parts of the earth. This is of considerable interest because it can be used both to test a theory and to work out practical values for g at the same time. It brings out such interesting things as the role of experimental apparatus. At the same time, it can be used for *testing* as well as for *estimating*.

In addition, there are other sorts of activities, such as 'articulating' a theory more precisely. This is different from making certain values of certain factual readings more exact. For example, with the Newtonian theory of gravitation, a good deal of work was put into working out the value of the gravitational constant. This articulates the theory in a certain quite different sense, which is easily understandable. One has the theory stating that the force of gravitation is proportional to certain factors, but the factor of proportionality was not known at first. A further piece of work of an experimental kind was required to determine it. This 'articulates' the law in making it more precise and definite. Such 'articulation' of a law is different from getting a numerical value, say, for densities.

Quantitative laws, like the law of falling bodies or laws of constants which vary from material to material, come under the same heading, and study of the constant articulates a theory more exactly. In normal science we may find further elaboration of a theory, of the kind that led to Coulomb's law, or investigations to find the law of cooling. In this connection, mention should also be made of applications. This, I think, is a vital heading, because again it is one that is overlooked by metascientists. It is easy enough to find examples from Newtonian dynamics. One can first of all elaborate the theory in the simple terms in which it was originally given, and then decide that it would be interesting to exploit it in certain other fields, such as in other media like water or air or different fluids, and in this way develop the science of hydrodynamics; or one can work out the theory of mechanics for resisting media and consider what would happen to projectiles if they were shot out in a resisting medium. This gives a vast amount of work of a normal kind which requires a great degree of originality, and is highly characteristic of ordinary science as it goes along from day to day. Kuhn claims that a new theory is not at stake here. The application of the original theory consists simply in the use of the theory in a specific field where subsidiary hypotheses may have to be introduced. But there is no revolution in theory involved. Many of the above activities are forms of *applied science*; indeed 'normal' science for Kuhn becomes virtually synonymous with applied science.

This is a rather cursory description of the kinds of activities that Kuhn refers to as going on in the day-to-day work of the scientist. I have skated over it rather fast because it is all set out in Kuhn, who puts it very much better than I have done in my summary account.

It is a pity, however, that Kuhn should have used the term 'normal' science for this, for it implies that revolutions are not part of normal science. Certainly I would regard the revolutions by which one great theory is replaced by another as 'normal' to science. But this is not, as we shall see, the objection to Kuhn raised by some interpreters of Popper.

'Normal' science as puzzle-solving and as daily rebellion

When Kuhn comes to summarise normal science, he describes it as puzzle-solving. He means that, in exploiting a theory in some new direction, you have a succession of knotty tasks calling for great ingenuity, even genius; and what you set yourself to do is to unravel the puzzle. Let us expand this. In dealing with a puzzle, until it has yielded there is some difficulty. Now a difficulty is either a *task* or a *problem*. When the puzzle does not yield to repeated expert attention, the difficulty is regarded not as a task to be accomplished but as a problem requiring some new solution. However, when a difficulty arises, you do not at once go about it by regarding it as a problem and scrapping the fundamental theory; you first try all the devices of puzzle-solving you can think of to accomplish the initial task.

This is the first big metascientific point that emerges from Kuhn and where he thinks he is differing from Popperian metascience, and where the followers of Popper also seem to think that Kuhn is differing from them. My first main point is that fundamentally there is no clash here at all, only an appearance of one which arises from the fairly simple oversight. To get at this we must go back over one or two metascientific matters.

Popper's fundamental metascientific thesis is that genuine scientists are always concerned to refute hypotheses, to test hypotheses by seeing if they can be refuted. Naturally they do not want to end up by having hypotheses or theories refuted; but they must be *capable* of being refuted if they are to be acceptable at all or scientific. (This, of course, is a very large-scale metascience which I am putting in a sentence.) Popper's metascience has then been applied to the history of science, but the application has been over-simplified or hasty. It is to the effect that all history of science (apart from the formation of theories) must consist of attempts to test theories (and therefore consist of attempts to refute theories)— that is so at all points all the time, *even* when the various kinds of activities described above as 'normal' science are engaged upon, even when checking a density point or applying a theory. In fact, it has been attributed to Popper that he has described the history of science as 'revolution in permanence'. This phrase has been used of recent times to describe the natural application to the history of science of Popper's metascience. I want to show that it is not a proper interpretation from the Popperian metascience, and therefore that there is no clash with Kuhn (in this area). The difference between the two can be put broadly like this. On the view accorded to Popper, every theory is all the time in all circumstances being tested, no matter what applications are going on. Thus the daily work of science, if not revolutionary, aims at small-scale rebellion (which might blossom into a revolution at any moment). According to Kuhn, the situation is quite different: scientists do not attempt to test a theory at all; in certain circumstances they are concerned simply to exploit it, use it, extend it, apply it. The doctrinaire Popperian might, to this account, reply, 'Of course, you may be extending the theory, but if the extension goes wrong, this is taken as a test of the theory.' It is this contention that requires modification; it is overstated — many other lines have to be pursued before the fundamental theory is questioned.

We are now in a position to round off the discussion of 'normal' science. In the phase in which a theory is being applied to certain fields and to some particular problems, or in which anomalies are met by exploring possible modifications in initial conditions, reconsidering the soundness of experiments, etc., the theory itself is not under fire. And this constitutes for Kuhn, very properly, a phase of 'normal' science. This phase *may* lead to what has been finely characterised by Lakatos as a form of research programme. Now that the nature of applications and the variety of modifications that may be introduced into the minor premisses have been considered, it is clear that this phase is wholly compatible with Popper's metascience. It is unnecessary and inappropriate to insist on a doctrinaire version of Popper, in which all scientific activity (applications, modifications of auxiliaries, etc.) consists of testing the main theory. Kuhn is right that 'normal' science is exploitation of fundamental theory and not a test of it; and this is also the proper application of Popper's metascience to the history of science.

In short, what I would call 'routine' science is further and wider applications, detailed corrections, and new methods of obtaining old results.

Notes

1 In addition, Agassi (1963) has attacked the idea of history as a continuous development in which any step forward can be found to contain transitional moves, and the idea of history as a quasi-automatic unfolding of new truth from preceding phases. Further, Agassi has

contributed two additional strong metascientific criticisms on the writing of history: he shows that most of it is based either on the inductivist approach or conventionalism, and he delivers a sustained attack on these as inadequate to the interpretation of history.

2 My student, Mr Stuart McCormack, has drawn my attention to a passage in Butterfield (1947, p.180), which I had missed, that is out of keeping with the 'key':

> So, as the seventeenth century proceeded, the minds of men had traversed and re-traversed the fields which we have been studying, putting things together this way and that, but never quite succeeding, though a certain high pressure was clearly being generated. One man might have grasped a strategic piece in the puzzle and, in a realm which at the time hardly seemed relevant, another scientist would have seized upon another piece, but neither had quite realised that if the two were put together they would be complementary. Already the scattered parts of the problem were beginning to converge, however, and the situation had become so ripe that one youth who made a comprehensive survey of the field and possessed great elasticity of mind, could shake the pieces into the proper pattern with the help of a few intuitions. These intuitions, indeed, were to be so simple in character that, once they had been achieved, any man might well ask himself why such matters had ever given any difficulty to the world.

It would seem that Butterfield marred his account of the 'key' and 'sluice-gates', but his view in the above quotation of how one finds the key was by putting the pieces of a jig-saw puzzle together, when the time was 'ripe'. This would require the services only of a plodding researcher with 'the help of a few intuitions'. Butterfield goes on to ruin his fine work by adding that these intuitions were so simple that one would wonder why there had been any difficulty. This is a travesty of Newton's insight and indeed of every major development.

3 Agassi (1966) has also pointed out the kinship of Butterfield and Kuhn.

References

Agassi, Joseph (1963), *Towards an Historiography of Science*, published by *History and Theory*, Beiheft 2, Mouton, The Hague.
Agassi, Joseph (1966), 'Revolutions in Science, Occasional or Permanent', *J. Hist. Philos., 3*, pp.47–61.
Butterfield, Herbert (1947), *Origins of Modern Science*, Cambridge.

Hattiangadi, J.N. (1965), 'Truth, Acceptance, and Agreement: a Discussion of Prof. Kuhn's Theory of Science', MA Thesis, University of London, Ch.V.

Kuhn, T.S. (1962), *The Structure of Scientific Revolutions*, University of Chicago Press, Chicago.

PART II: OBSERVATIONALISM

9 Observations *not* the building blocks of science: the structure of science as theories of layered observations

The relation of data to theory may assume a different hue if we reconsider the nature of observation itself. For, common both to observation-manipulation, which is a functionalist approach to observation, and to inductive observationalism, which is a structuralist use of observation, is the ultimacy of observation. In both, observation takes precedence over theory, in that if there is any tension between them it is theory that always and necessarily has to give way. Theory has run no risk for 400 years of being a sacred cow.

In this sense observationalism has been for centuries and still is at the root of scientists' (I do not say scientific) outlook.[1]

This impregnable position bestowed upon observation may be questioned on two counts. (i) There is the simple fact that at times, even though not often, an observation has to give way to a theory: that is, at times an observation has to be questioned and even rejected in favour of a theory. This must be known to most working scientists, but its metascientific significance is usually overlooked. (ii) There is a much more subtle situation that arises if we are correct in supposing, as Popper (1959) asserted, that every so-called observation is theory-laden or permeated with interpretation, or, otherwise put, that there is no such thing as a 'pure' observation, an objective datum independent of all experience, as it were waiting to be photographed by a human mind.

With many examples it is obvious that unsuspected interpretation is present. Is this general? It seems to me that an observation is a selection formed in terms of certain ideas the observer brings to bear on the world. There would possibly be little tendency to dispute this, were it not for one point that constantly obtrudes: how can there be interpretation unless there is something uninterpreted in the sense of being something upon which interpretation$_1$ is imposed, although in itself it is an interpretation$_2$ of . . . etc.? This smacks of artificiality, for there seems to be no reason for the dual role played by any particular layer; also the sequence seems to be an actual infinity present before us. However, it can be altered to the following: an ordinary observation$_1$ is an interpretation$_1$ from which we can *abstract* what is, in relation to *it*, a pure observation$_1$; this abstraction is in its turn seen to be an ordinary observation$_2$ in a different context, that is, to be an interpretation$_2$ from which we can again abstract what is, in relation to *it* a pure observation$_2$; and so on. In this case the layering goes on only so long as we make it go on, that is, by making abstractions.

Now there are unexpected consequences of there being no ultimate uninterpreted observation. What effect does it have on the dominance of observation over theory?

With instrumentalism, we should be pushed into the contention that a theory is an instrument for computing entities that are precluded from being pure observations and that are therefore a fusion of instrument and observation, or rather an observation interpreted as an instrument. About this it is to be noted that an instrument for computation is not an entity but an operator; hence if it calculates only another instrument, then it is calculating an operator and not an observation in the sense of something that is the case. Hence instrumentalism cannot be articulated on the basis that an observation is itself a blend with an instrumental component.

Again with conventionalism, we should be making a theory or convention fit not an observation but a fusion of which one component is a convention. Now a convention to make observations tractable is one thing; a convention for handling conventions is another. A convention describes not an entity but an operation. Hence, as with instrumentalism, conventionalism cannot be articulated.[2]

Turning to induction, an observational premiss has to be regarded not as pure but as a compound embodying a theoretical component not yet justified by induction. Thus an induction cannot be justified from its premisses without justification of its premisses which require inductive justification. Hence inductive justification presupposes an infinite regress of inductive justification.

Thus a common element of all the functional as well as the structural approaches is the ultimacy of observation. But we may note that if this is untenable the dominance of observation over theory must go.

It is widely noticed that observation is theory-impregnated, but its full significance seems to be hardly recognised. So it is a matter of some interest and importance to ask what difference it makes to science itself what interpretation is adopted. Consequences of observationalism are:

(i) The shared defeatism about knowledge of both instrumentalism and conventionalism, that is, the acceptance that reality is unknowable, obliterates the distinction between theoretical science and technology on the instrumentalist and conventionalist interpretations — theory becomes pencil-and-paper technology.

(ii) In a clash between theory and observation, observation *must* always win.

(iii) If a theory cannot in principle represent reality, the (?insoluble) problem arises of how a theory (instrument-convention) can function — can make predictions/patterns of observations (which would not be a problem if a theory represents reality, that is, is connected with observations in accordance with it).

(iv) For induction, theories become algorithms. While some algorithms are ingeniously contrived empirically, it is common for them to be derived from theories (presumed not to be algorithms); hence no non-algorithmic theory would be available for providing algorithms. This would seem to be so even for instrumentalism and conventionalism.

(v) These are meta-scientific considerations. Are there any that bear on the working scientist? More specifically, are there any questions he cannot ask or solutions he cannot use? It would have been improper to postulate the atom and (after agreeing that chemistry could proceed satisfactorily without it, simply working 'as if' it existed) it would have been improper to look for it; it would have been improper to postulate 'attractive force', or the electron — our knowledge not only of the world but of science would have been wrongly impoverished. And it is wholly obscure, on these positions, why scientists get so excited about science (Synge, 1970).

The position, then, is this: if observations are theory-laden, and if theories are solely instruments, then observations are instrument-laden. Hence the task of science cannot be to find a theory (or instrument) to pass from a pure observation (or non-instrument) to another one.

With conventionalism, if observations are theory-laden, and if theories are solely conventions, then observations are convention-laden. Hence the task of science cannot be to find a theory (or convention) to pass from a pure observation (free of convention) to another one.

Part of this is applicable to induction. Induction requires premises to be purely observational, uninterpreted; for an interpreted observation would be theory-laden. But if all observations are interpreted, we should lose the characteristic point of induction, of getting theory from

73

pure observation.

This argument, if effective, would refute induction both as inference and as method, but it cannot do so without aid. It could, however, be put in the form of a dilemma of choice: you cannot accept both induction (inference or method) and observation as theory-laden. If observations are theory-laden, the approach collapses. While the actual logical procedure might be satisfactorily transformed into a version involving a benign infinite regress of inductive justification, the raison d'être of the method, that is, the classical empiricist demand for a firm root of experience, would be lost. Thus the problem of the gap between data and theory is transformed by the assumption that observation itself is a blend of theory and observation.

Appendix: Science as a Form of Currency Reserve

Induction is the vault of the scientific (computerised) bank: the vault stocks huge reserves of observations/facts/information; the computer processes this raw material into various forms of currency, according to whether the customer is a pure scientist, an applied scientist, or a technologist. These customers not only make withdrawals of processed currency; they also make huge deposits (of observations/facts/information).

Instrumentalism is a central clearing house, which governs transfers of observations/facts/information from one bank to another.

Conventionalism is the world-treasury which decides whether to inflate or deflate, whether to revalue the currency, whether to modify its purchasing power — by which the reserves have whatever value it ascribes to them.

Might it not be that these vaults are as useless as Berkeley claimed were the vaults of Amsterdam, full of unused gold? Might it not be that there are other vaults, not known to many metascientists, small in size, stocked with a few (maybe a few dozen) theories? That a pure scientist carries some of them in his wallet in the form of film which he looks through when he studies nature? And that sometimes a technologist takes one of these to a shop? And that occasionally a scientist makes a film from something in his mind's eye, and deposits it in the central bank?

Notes

1 For some very trenchant criticisms of observationalism interpreted in terms of sense-data, see Skolimowski (1969). The significance of

observationalism was appreciated by Feyerabend (1962) and Agassi (1966).

2 Likewise with operationalism: operations would not be pure but would be fusions consisting of operations permeated by theories that inevitably contain concepts not yet operationally defined. Thus operationalism cannot be formulated without infinite regress.

References

Agassi, J. (1966), 'Sensationalism', *Mind,* 75, pp.1—24.
Feyerabend, P.K. (1962), 'Explanation, Reduction, and Empiricism', in *Minnesota Studies in the Philosophy of Science*, Vol.3, ed. H. Feigl and G. Maxwell, University of Minnesota Press, Minneapolis, pp.28—97.
Popper, K.R. (1959), *The Logic of Scientific Discovery*, Hutchinson, London and New York.
Skolimowski, H. (1969), 'Knowledge, Language, and Rationality', in *Boston Studies in the Philosophy of Science*, Vol.4, Humanities Press, New York, pp.184—5.
Synge, J.L. (1970), personal communication.

10 The metaphysics of observationalism: the heir to essentialism

On reaching the end of Part I of this book, it is desirable to state its claims. I have followed Popper in the view that the Renaissance emancipated modern science by repudiating essentialism. I have added that in 400 years of modern science, essentialism has been replaced by observationalism (and its handmaid induction) both by philosophers and by scientists. Observations on this view, when correctly made, are true and incorrigible — and induction arrives at generalisations that are true and incorrigible. Interestingly enough both observationalism and induction share with essentialism the components that the notion of physical reality must be taken seriously and so must the notion of truth, handmaid of reality, in that truth is obtainable and reality accessible. Instrumentalism and conventionalism presuppose observationalism, though not induction, and thus, while admitting observational truth, deny both truth and reality at the theoretical level. Or rather, instrumentalism has no need of the concept of 'truth', and conventionalism sits on the fence about it. The role of 'truth' in scientific inference, however, annihilates instrumentalism and seriously weakens conventionalism.

They may all be seen, further, as responses to the observation—theory gap: how to get to theory from observations; answered by the inductive, instrumentalist, and conventionalist strategies of observation-manipulation.

Thus the philosophy of observationalism has been and continues to be a basic presupposition of these interpretations of the nature of science.

When twentieth-century scientists tried out instrumentalism and conventionalism in place of induction, they took no risks over essentialism. For they retained observationalism, and gave up the notions of 'truth' and 'reality', which are vital to essentialism. Thus the new interpretations of science, namely instrumentalism and conventionalism, are strongly anti-essentialist. The choice made is perhaps an indication of a need to ward off essentialism — of over-reacting.

There was, after all, an alternative: observationalism could have been questioned. It is now put in question by the notion that observations are theory-laden. What would have been the threat, or what is now the threat, from such a notion?

We get a hint at the answer by considering the subsequent course of rational enquiry. It was the empiricists and main contributors to natural science who fastened on observationalism. The opposite school of intellectualists distrusted observation and put their faith in the a priori and thus in theory. And they were to a man essentialists. What this suggests is that the empiricist philosophers and their scientific relatives would have feared emphasis on theory because of the threat (actually a mythical threat) of essentialism. Hence they could not have taken up the notion that observations might be partially theoretical.

Further, the distrust of theory that pervades the world of natural science — even though theory is its life-blood — is very clearly manifested in instrumentalism and conventionalism, though more strongly in the former. Now the need for certitude is one obvious root of this attitude. But the attraction of certitude was not enough to seduce the empiricists towards espousing theory. So it looks as though some other distraction weighed more fully. And the obvious possibility is that theory, taken seriously, claims to tell us about reality. Now this is not in itself necessarily essentialistic in its implication — but it is easily read that way by anyone looking over his shoulder at essentialism.

Thus in addition to Popper's contention that modern science became emancipated by repudiating essentialism, the further contention would seem to be required that science is still reacting against that philosophy by clinging to observationalism.

Other trends in science work out in a surprising way.

A respect in which natural science, even into the twentieth century, evinces involvement with essentialism lies in its overall strategic method. Taking the method to be hypothetico-deductive — when it may or may not be interpreted essentialistically — we note in addition that the framework of science, its deepest weltanschauung, is its reliance on mechanism, or rather what I have called *inert mechanism*.

There have been considerable changes in the overall scientific framework during the past four hundred years since Copernicus, and it may be instructive to list some of the variations that are relatively specific:

'Action by contact' replaced by
'Action at a distance' replaced by
'Action by contact'.

'Nature abhors a vacuum'.

'Nothing comes of nothing' replaced by
'Spontaneous creation of matter'.

'Indivisibility of matter'.

'Compounds are composed of elements in simple proportions'.
'Science is built upon measurement/experiment/observation'.

Here, however, what is of concern is very general. Thus:

'For all *physical* changes seek only *physical* causes.'
'For all *physiological* changes seek only *physical* causes.'
'All *physical* changes/natural phenomena are wrought by *mechanical* process.'[1]

The first of these expresses the naturalism of the new era, but does not fully represent the activity of natural science. The second, which is a slightly more determinate prescription, reflects the professional attitude in all physiological research; it has yielded colossal dividends, even if it has sterilised other research that might also yield significant results. In expressing organicism it also expresses epiphenomenalism (which no physiologist believes in once he has left his laboratory) and, as we shall see, a reductionist approach. The last one, which Marie Boas (1952) and Rupert Hall (1962) have aptly called the mechanical philosophy (or the corpuscular philosophy, which rather confines it to physics), expresses the notion of mechanical process, or mechanism, which in general characterises the spirit of natural science.[2] For a long time it did characterise most of natural science in fact, e.g. Galilean kinematics, Newtonian dynamics, and so on (see also Popper, 1958), but it does not literally characterise Newtonian celestial mechanics or electricity or magnetism. (Newton may not have looked on himself as a mechanist, but the Newtonians did.) This is easily lost sight of because these areas were somehow subsumed under the outlook of dynamics. It has been implicitly recognised, on the other hand, by attempts to eliminate action at a distance from celestial mechanics. Such matters, however, are hardly germane: the main point is the conception of Nature as *clockwork*, the broad conception of *mechanical process*. It serves most purposes to refer to this outlook briefly as 'mechanism'. Parallels in the social science area of psychology will be given below.

Mechanism has been a dominant part of the weltanschauung of natural science for four hundred years. Whatever its deficiencies, in its debt are all the theoretical achievements that have accrued ever since. It may seem too trivial to prate about, but the scientific revolution that began in 1543 was just this — the replacement of theocentric/anthropocentric astronomy and physics by mechanism. Before that time God was regarded as a clockmaker who made and looked after his clock, Nature. Afterwards physics contained scarcely a reference to clock-maintenance, though the conception persisted in philosophical concomitants. God continued to have a role, underpinning the laws of physics, but the role was restricted to this and did not include the detailed management of events; secularism meant just this 'dominion status' accorded to the physical world. Mechanism originated in secularism and has been with us ever since (Wisdom, 1971).

Departures from the framework of inert mechanism are always strongly resisted. Departures have been made for nearly fifty years now, e.g. in psychosomatic medicine. But, although the stronghold has been breached, hard-line loyalists have rallied round. The interesting point is that nothing in science requires us to postulate inert mechanism; but it has been deeply ingrained in the minds of natural scientists for centuries. The reluctance to work the hypothetico-deductive strategy without it seems to show rather clearly that inert mechanism is regarded as of the essence of science. Thus in this regard science strongly turns out to be essentialistic.

And it is essentialistic over observationalism: in the beginning was the observation and without it was not any theory made that was made; from observation science begins and to it it returns. Observations are the reliable incorrigible building blocks. It seems clear, however, that the trust placed in observation is due not to its substantial position in the universe but to its metascientific role in knowledge, certainty, and opposition to inner essences.

On this account science replaced essences of substances by essences of method. But the stress on the method that involves inert mechanism surely protects science from an ontology of ghosts and goblins, vampires and witches (cf. Wisdom, 1968).

Thus the survival of essentialism in science, in its method, has the function of repudiating essence as regards substances.

Notes

1 A wholly satisfactory terminology is difficult to find; a good suggestion is 'mechanical process', which I owe to Mr J. Robert Bath of Fredonia, N.Y.

2 'The essence of mechanistic philosophy in the seventeenth century was the axiom that all natural phenomena could be reduced ... to one single kind of change, the motion of matter' (Hall, 1962, p.101).

References

Boas, Marie (1952), 'The Establishment of the Mechanical Philosophy', *Osiris, 10*, pp.412—541.
Hall, A.R. (1962), *The Scientific Revolution*, London, pp.205—16.
Popper, K.R. (1958), 'Philosophy and Physics', *Acts of XII Inter. Congr. Philosophy*, Venice, Vol.2, pp.367—74.
Wisdom, J.O. (1968), 'Anti-Dualist Outlook and Social Enquiry', *Proc. Inter. Colloquium in the Philosophy of Science*, ed. I. Lakatos and A. Musgrave, North-Holland, Amsterdam, pp.433—9.
Wisdom, J.O. (1971), 'Science versus the Scientific Revolution', *Philosophy of the Social Sciences, 1*, pp.123—44.

11 After observationalism — whither?

With modern science risen from the ashes of essentialism, and provided with a new foundation in the philosophy of observationalism, we have confronted the weaknesses of conventionalism, which was the most important interpretation of science centred on observationalism, and we have confronted deficiencies in observationalism itself. Moreover, we have seen modern science risen more generally from the ashes of metaphysics, and yet we have seen that, with the basic role of 'truth' in science and its ontological cognate 'reality', metaphysics has reappeared.

Metaphysics shows itself in three places: in the thesis that observations are theory-laden; in the problem of 'reality', concerning what constituents of the universe are 'ultimate'; and in the role of metaphysics in governing scientific theory. In opposition to conventionalism and observationalism, Popper's metascience will be the springboard for Part II, though it gives no indication of the landing point. This is partly because his metascience eschewed metaphysics, though more than its rivals it shows the need for metaphysics. In comparison with the traditional philosophies of science, Popper's operates differently. By tackling the demarcation problem of how to characterise the difference between science and non-science, Popper produced his falsifiability-criterion for empirical content. Popper himself, without particularly going into the matter, has always taken a 'realist' view that scientific theories tell us about the world. This, however, runs into difficulties of its own: once it is realised that for Popper falsification is not absolute — neither in practice nor in principle. For this reason it was necessary, especially in view of Kuhn's resuscitation of conventionalism (to say nothing of misinterpretation of Popper's position), to develop Popper's view more fully over the matter of refutability. Yet even with such a straightening out of the falsifiability-criterion of what constitutes scientific (empirical) content, the difficulty remains of how to support the realist view of the world, if refutation of a theory is not capable of being carried through; for it leaves a loophole for the conventionalist interpretation.

On top of this is another consequence of Popper's metascience: that observations are theory-laden. This putative conclusion has the enormous advantage of destroying observationalism (which holds that observations can exist and be noted uninterpreted), and also destroys induction and instrumentalism. It still further weakens conventionalism, without however destroying it, thus leaving Popper's metascience with a rival, however weak. Moreover, the very fact that observations are theory-laden, that is, are interpretations, indicates at least the possibility that observations do not depict reality. (And the bogey of scepticism is always lurking in the shadows.) It is therefore a fundamental issue whether conventionalism can survive the various forms of undermining that have been argued in Part I, or whether realism, in some form or other, can be found to be viable. This formidable issue is in a significant sense metaphysical.

If modern science took its start and its inspiration from discarding the philosophy of essentialism 400 years ago, it rested negatively upon a rejected metaphysics. But it developed over 400 years positively upon the philosophy of observationalism — itself a metaphysic hoisted over the rejected metaphysic of essentialism. Is contemporary science to remain founded on this presumption of observationalism? Is it that the Popperian alternative cannot be rationally supported?

In attempting to deal in Part II with the issue between conventionalism and some form of realistic alternative, we shall find that it turns on a presumption, which is also metaphysical. But then, not only is there a difficulty on the Popperian metascience about the question of what is real; we discover also that metaphysics, which Popper's demarcation criterion placed outside science, turns out to be intrinsic to science itself, and, more serious still, the metaphysical ingredient is not amenable to Popper's testability criterion for empirical content. We are then fully launched upon the enquiry about the role and kinds of metaphysics in natural science.

12 Observations as theory-laden

For several hundred years it has been maintained by empiricist philosophies that observations are 'pure'. What is conveyed seems to be a model in which an observation is, so to speak, a mask painted on the world which is there unseen until our eyes open like the shutter of a camera; and an event, consisting of seeing, occurs inside the observer. Some such notion has persisted in classical empiricism from the time of Bacon until the present century — that is, within empiricist circles. It was, of course, repudiated entirely by all schools of idealism; it was not accepted by Kant; it was quite foreign to the outlook of the intentionality school of thinkers in Austria and of their successors the phenomenologists. The first empiricists to dispute it were Popper (1934), who held even in his first book that there is no such animal as a pure observation, and simultaneously Ajdukiewicz (1934), who according to Skolimowski (1967) denied any real difference between report-statements and interpretative statements. This challenge took a long time to be accepted in empiricist circles but it became accepted quite widely in the post-World War II school of language analysis in Britain. In this last school is assumed the specific form that sense-data are in effect a myth. In the philosophy of perception, which occupied British empiricism during the twentieth century, sense-data were regarded as pure observations in the classical sense — a view that had a run of about fifty years. The wide implications of such a change did not attract much notice for a long time, but much began to be made of it after 1962, when Kuhn and Feyerabend either implicitly or overtly spread the word in philosophy of science. (The use to which they put it, which in effect was to introduce a note of irrationalism into science at its most extreme and which at its least extreme was to introduce or reintroduce conventionalism, is not a matter for this chapter.) I think in fact that the significance of the new doctrine has not yet been completely explored. My purpose here is to consider the doctrine and to raise the question how far it can be established or what reasons can be advanced in support of it.[1]

At first sight the notion of a theory-laden observation would seem to be an absurdity, for observations are needed according to empiricist theory in order to build up knowledge, and although knowledge may be doubtful, at least with observations we have got some firm beginning. And moreover we need pure observations with which to test the theories we arrive at. On the other hand, if observations are heavily theory-laden then we are, as it were, on a quicksand trying to build up theories on observations that are already full of theory. This kind of difficulty, however, is not hard to counter or to overcome, because it is well known that theories exist at a different level of abstraction, and clearly the theory that is embedded in a theory-laden observation is a very low level of theory.

Let us now consider a fairly wide range of examples to see at least how far the notion of a theory-laden observation has any measure of plausibility.

In the higher reaches of physics there is no doubt about the matter at all: observations are full of theory and every physicist who reflects about the matter at all knows it. Thus if we take a Wilson cloud chamber, which is a very sophisticated piece of apparatus, and observe the tracks that are visible, the observation is of a subatomic particle travelling through the chamber; but this is not an observation in the classical sense at all, for all one can see is a line consisting of steam particles or a line consisting of a track of steam. This example, however, belongs to very advanced physics and is far removed from more modest ranges in physics, let alone everyday experience. So let us consider readings on a volt meter. We wish to know whether our car battery needs to be recharged, so we measure its charge with the volt meter, but the reading on a volt meter is one which is thoroughly impregnated with electrical theory, so we are obliged to try elsewhere for a theory-neutral observation.

The patient seems feverish and puts a thermometer in his mouth. The mercury column in it rises and he notes that it rises above 36.9, whereupon he reaches for a bottle of aspirin. The reading of a thermometer depends upon theoretical matters to do with the expansion of mercury as between ice-cold water and boiling water (and indeed that is not the end of the matter, as we shall see in the next example).

Let us endeavour to do something simpler still. Let us measure the height of a mountain. Surely the height of a mountain does not involve theory. Mt Everest is (was) 29,002 feet high and if we have measured it correctly that is the height we obtain. But let us reflect what height means in connection with a land mass; it means height above sea-level. Now what is sea-level? Sea-level is an arbitrary point selected for convenience as an average between high and low tides, which indeed varies from time to time during the year. So the height of the mountain is an arbitrary measure that depends upon a statistical smoothing out of figures measuring high tides and low tides.

Take a still simpler example. We tap the barometer in the hall to see whether the day is going to be fine or whether there will be rain. We note the atmospheric pressure shown on the barometer. If, however, we work in a physics laboratory this is not good enough. We need to know the atmospheric pressure at NTP, that is to say at normal temperature and pressure at sea-level. So here we see once again that the atmospheric pressure, which would seem to be a reliable observation, is an arbitrary figure depending upon theories not only of what sea-level consists of but also of the behaviour of barometers as they move from sea-level to the site of our laboratory or our hall.

So in physics 'observations' are observations no one makes. They are very often calculations, and complicated calculations at that, which depend both on mathematics and upon physical theory. It is clear that we are not going to get very far in physics in our search for pure observations. Let us therefore not waste time on a lost cause by searching for pure observations in physics but turn to matters of commonsense for examples. We observe a piece of ice melting in warm water. Melting, we say. Is this a theory-neutral observation? In fact it is infused with a long line of theory that ice consists of the same basic material as water and is transformable into it.

Let us try a simpler example. Consider the observation that a cork is floating in a bowl of water. Floating? Does this mean that the cork has been suspended in the water by Zeus? Nowadays we think differently. We think that it is due to an Archimedean thrust. Either way it involves a theory. In desperation we try again and look for something much more mundane. For example, we see a dog running after a cat. But let us reflect: we have such a strange idea that, if dog is running in a certain direction in line with cat and cat is running further away in the same direction, dog is trying to catch cat. Dog might be simply trying to show cat he could run faster. Or if dog were English he might be showing the propriety of forming a queue.

We may as well admit that all substantive thing-terms have wider implications than what 'meets the eye', which is simply the front side. So the thesis that observations are theory-laden is hard to dispute with regard to *objects*. Is there any realm of observation at all in terms of which we might try to challenge it?

Our only chance would seem to be in the case of so-called secondary qualities such as red or round, or sense-data such as the observation that a curve that one is inspecting is a closed one. Now this total observation of the curve and its shape hinges on the sense-datum involved, namely the closed curve. So 'I see a round shape' is, so to speak, the last ditch; surely here we have a pure observation completely free of being theory-laden.

Now at last we begin to see the issue opening up. We ask ourselves whether this observation is possibly a pure one, and we realise that the reason for supposing that it is lies in the classical notion of a sense-datum, that is to say, something *purely private* to the observer making the observation. Now arises the challenge. Is this what is required of an observation at all? For an observation, not only in science but for any purpose, is one that must be 'intersubjectively open to inspection'. That is to say, *an observation would have to be intersubjectively open to inspection if it is to form part of the world* (cf. Musgrave, 1983).

If this were not so, if we did not wish to accept this conclusion we would have to maintain the alternative: that what is not open to public inspection is part of the world; in other words, that what is private to the perception of an observer is nonetheless part of the world. And obviously an observation in this sense has no place in science, for in science an observation has to be open to public inspection. About this last point we might even add that it is itself a theory that 'observation is invariant as between observers'.

We are here distinguishing between our *way of seeing* things (subjective) and what we *ascribe* to the world. Naturally it is not denied that there is such a thing as a private perception nor that this may be characterised as our *way* of perceiving, while nonetheless *what* we perceive belongs to the world. Going back to our example about the closed curve, assuming that it is not a private observation peculiar to one observer but is an observation that may be relevant to science, then *this observation is infused with the theory that it is a closed curve for all observers*.

The result of this discussion may be put in the form of an interesting dilemma: a pure observation would ipso facto be incorrigible and this would be possible only if it were possessed by one observer alone; for, if it were accessible to more than one observer, the incorrigibility would be incompatible with the possibility in principle of another observer seeing it differently from the first one. Thus, the question of pure observation versus theory-laden observation becomes inherently linked with the question whether there is one observer or several. The linkage effect is such that a pure observation would belong with a solipsistic view, while the thesis that observations are not pure, that is, are theory-laden, requires the acceptance of the existence of other observers. The observer has to make the assumption that others like himself exist, that is, 'I see what they see'. Now this is not an analytic proposition: that is to say, it is corrigible, that is, theory-infused. In short, incorrigibility and pure observations imply that I alone exist — a result that is an ad hominen (!) absurdity. The dilemma in question thus comes to this: you either accept that observations are theory-laden or you accept solipsism. (Russell, in basing knowledge on pure observation, had a wholesome respect for the threat of solipsism but admitted an inability to cope with it.)

Now it is important to point out that this is not a proof of the claim that observations are theory-laden. For if it were, it would be a proof also that other observers exist. And no such strong proof is claimed or is needed: it cannot be claimed because it would be equivalent to asserting a synthetic a priori proposition (which I have, I believe (Wisdom, 1975), shown to be neither proved nor disproved). Disproof of pure observations as the foundation of all knowledge would be a proof of the existence of other observers, which is impossible; hence the contention that observations are theory-laden, or, otherwise expressed, are interpretations, has the same hypothetical status as the existence of other observers (to which it should perhaps be added that the claim for observations that they are theory-laden implies that other observers exist, but that the converse does not hold — it is not the case that the existence of other observers implies that observations are theory-laden; the relationship is not that of a biconditional). But if we wish to follow a usable working hypothesis for science and other purposes, then we can equally use the working hypothesis that other observers exist and the working hypothesis, which is its concomitant, that observations are theory-laden.

The dilemma may be expressed in the form of a theorem: the assertion that observations are not theory-laden implies the assertion of solipsism.

Appendix 1: The Fallibility of Observation — A Medical Example
Alan Thompson[1] and J.O. Wisdom

The idea that observations are theory-impregnated is now widely accepted. There is, however, a distinction to be drawn between an observation that is *impregnated* by a theory and an observation that is *altered* by a theory. The following observation is offered not to add to the argument over the issue but simply as a striking specimen of the latter.

At a clinico-pathological case-conference a patient's illness, from which he had died, was presented for discussion of diagnosis and pathogenesis (the autopsy results were withheld).

The course of the fatal illness was described in some detail by the pathologist, including comments made during the illness by various medical authorities, none of whom joined in the discussion.

Of the several observations made about the patient and his illness various explanations were offered, some of which could be readily ruled out. They fitted in, however, with a well-recognised syndrome, consistent with the diagnosis of a brain tumour at a specific site — with the exception of one observation, namely, that of a paralysis of both sides of the face. A bilateral facial paralysis could not be explained if the condition was in fact a tumour in the suggested site (for this would not affect the facial nerves). It was an inexplicable feature obtruding in an otherwise orderly sequence of events in a defined area of the brain.

If, however, this inconsistent observation of 'paralysis of both sides of the face' were to be replaced by the observation 'immobility of both sides of the face' this modified observation (a) would probably be an accurate description of what was actually seen and (b) would be consistent with the suggested diagnosis; it would indeed even support and amplify such a diagnosis, for it would supply yet another localising feature which would be explained by a tumour at the supposed site.

What this amounted to was the suspicion (a conjecture) that the 'observation of paralysis' was not indeed strictly an observation, but rather a hypothesis (interpretation, conjecture) about the nature of the immobility. Accepted at its face value (!), what might have been taken to be passive immobility was taken to be paralysis, so that the case-history allowed of no satisfactory diagnosis; accepted as a debatable interpretation, that is, that the immobility was possibly not a paralysis, the case-history could be diagnosed as due to a tumour, and this would be a consistent hypothesis.

The autopsy revealed the existence of the tumour at the site conjectured.

This necessitated correcting the case observations, by discarding that of facial paralysis on both sides of the face and replacing it with an observation of immobility. The bilateral immobility would support the diagnosis because the supposed tumour was located in that part of the brain where the spread of damage may lead indirectly to a state of facial immobility — the so-called Parkinsonian 'mask' — which might be mistaken for facial paralysis (especially if the possibility of Parkinsonian mask is not thought of).

From this little episode it is clear that an observation can be questioned and can be successfully questioned. The basis on which it is questioned is the diagnosis, that is to say a conjecture or a theory. In such a situation, contrary to tradition, theory takes precedence over observation. It is, of course, then subject to observational check. But having received the observational check, the theory is vindicated and the original observation quashed.

Some further elucidation is needed about the nature of the original record of the observation. Something about the record was correct: there was an immobility on both sides. But it is interesting to note that a medical hypothesis (interpretation, conjecture) was at once read into the observation, or embedded in it, and recorded as part of the observation itself. For the observation was recorded not as an immobility but as a paralysis. Now one may be able to observe an immobility, but one cannot observe a paralysis — that is a hypothesis (interpretation, conjecture). Whether or not there is such a thing as a pure observation without an interpretative element embedded in it, it is obvious that many observations are loaded with interpretations and this alone affords justification for querying an observation.

Loaded observation-reports are very common in medicine. Practitioners do not usually report 'primary' observations (if indeed there are such things at all), but rather read interpretations into them — in terms of paralysis, infection, murmurs, and so on — and it would be excessively tedious to do otherwise, not to say misleading in many instances. But it does leave the door open for faulty interpretation being accepted as 'primary fact'. The saving clause often is the inconsistency of the one so-called observation which sticks out like a sore thumb. One may have to discard it to get a consistent diagnostic picture, and in view of the rather random nature of clinical observation, its lack of precise measurement, and its often hasty and incomplete nature, such discarding, even though tentative while shaping a hypothesis into a diagnosis, is common enough.

Here we have a striking illustration of an observation being open to challenge, having to defer to a theoretical idea, and being in however modest a way itself theoretical in nature.

Appendix 2: Die Philosophiedämmerung

This is perhaps an appropriate place to add a note on Feyerabend and on the similarity of some of his recent views and on the great difference between his final conclusions and the 'overall view' of this book. I had already made occasional (approving) references to his earlier writings, but no references to his later work.

His views got around as slogans — 'anything goes', 'anarchism', and one could add 'rationality to the guillotine', 'all wild ideas (including myth and superstition) acceptable'. Reviews did not moderate this impact, so I didn't suppose one could take him seriously, and I didn't read his later work. At my publisher's request I decided as a chore to try Feyerabend's (1975) *Against Method*. I read nearly all of it rather carefully towards the end of the summer of 1980. I found a pot-pourri. I found nearly all of the first half acceptable and similar to my book in its attack on observationalism (not that he would have owed me anything, as those parts of my book that had appeared earlier in remote places would have been unlikely to come his way). While Feyerabend himself writes with sobriety and makes little use of his slogans, he does tend, however, to be rhetorical where I have made an effort to produce some argument in my handling of the subject of observationalism.

In the first half of the book, Feyerabend fires some heavy missiles at old myths; and he does a good job of criticism, although his lumping Popper with all the other weak brethren seems to me ill-judged. While advocating freedom from traditional intellectual procedures, he himself uses logical argument and rationality extensively at a commonsense level. In the second half of the book, however, Feyerabend indicates his willingness to listen to numskulls, half-wits, soothsayers, warlocks, and the like for profitable ideas — which is an abandonment of rationality. This is an admission of despair: for Feyerabend there is *no intellectual hope.* He holds that our capacity to be productive is rooted in anarchism. Not even logical positivism has gone so far in bringing about *die Dämmerung der Philosophie.*

Note

1 The late Professor Alan Thompson, FRCPI, was professor of medicine at the Royal College of Surgeons, Dublin.

Note

1 Musgrave (1983) has written a highly important paper on this subject; he reinforces theory-ladenness.

References

Ajdukiewicz, Kazimierz (1932), 'Das Weltbild und die Begriffsapparatur', *Erkenntnis, 4.*
Feyerabend, Paul (1975), *Against Method*, NLB, London; (1978), Verso, New York.

Feyerabend, Paul (1978), 'From Incompetent Professionalism to Professional Incompetence — The Rise of a New Breed of Intellectuals', *Philosophy of the Social Sciences, 8*, pp.37—53.

Musgrave, Alan (1983), 'Theory-Ladenness and Incommensurability', *Philosophica, 31*, pp.45—62.

Popper, K.R. (1934), *Logik der Forschung*, Vienna.

Skolimowski, Henryk (1967), *Polish Analytical Philosophy*, Routledge & Kegan Paul, London.

Wisdom, J.O. (1975), *Philosophy and its Place in our Culture*, Gordon & Breach, New York and London, Ch.10.

13 The conception of 'theory' as 'non-instantiative'

Certain substantive issues are involved in how the term 'theory' is used. Although the term 'theory' has become more or less standard both in science and in metascience (see Kaplan (1964) for a useful statement), there are still certain divergencies in its use. These divergencies are not simply an uninteresting matter of ambiguity in which the term is used to refer to different levels or types of entity; for substantive issues are engaged that concern the description of real though unobservable entities or the attempt to 'reduce' theoretical truth to what is manifest.

It is widely agreed that in science a theory is in some sense or other more general than a mere regularity, usually called a generalisation. The broad relationship between 'theory' and 'generalisation' is that generalisations are mathematical or logical consequences derived from theories either by strict deduction alone or by a combination of deduction and approximation. The direction of the deduction is a one-way traffic even where no approximation is involved, for it is not usually possible to derive a theory from a generalisation or any set of them. The theory is then said to explain the generalisations or the observations that come under the generalisations, that is, the observations that constitute the generalisations. A further requirement usually agreed upon is that the theory shall be capable of yielding a further generalisation other than the ones it was designed to explain in the first case: that is to say, the theory should have wider, unexpected consequences beyond those it was intended to cover. All this is generally accepted and it may be mildly surprising that there should be any peripheral matters giving rise to questions of any significance.

How much does 'theory' include?

It is generally agreed that a theory may consist of one hypothesis or it may be a set of hypotheses. In theoretical explanation the component hypothesis or hypotheses are universal; and, intuitively, most scientists or metascientists would restrict 'theory' to universal hypotheses. What then would they be excluding? They would be excluding statements of initial conditions and boundary conditions which are needed and brought in when it is a question of putting a theory to use. Thus if we wish to derive Kepler's laws of planetary motion from Newton's theory of gravitation, we shall have to insert an initial condition stating, for example, the velocity of the planet; the combination of these premisses, plus other auxiliary hypotheses which may be needed and taken from some other branch of the subject and assumed in this context without particular examination, is sufficient for the (approximate) deduction of Kepler's laws, that is, to enable his generalisations to be derived. If, however, we wish to discuss the motion of a comet the initial conditions will be different; and if we wish to discuss the motion of the tides the initial conditions will be again different. Thus the theory may be combined with any one of an infinite set of initial conditions, depending on the use to which the theory is to be put. And we may note that the use to which the theory may be put in such a context may be either one of two types: it may be explanatory scientific application to understand some generalisation or it may be a metascientific use to enquire whether or not the theory is testable at all.

A metascientist might conceivably be tempted to include within the term 'theory' not only the universal hypothesis but the infinite set of possible initial conditions. If he were to take this line he would have to allow that many of the possible initial conditions were totally unknown. It may not be easy to see what would tempt anyone to take this line but possibly the metascientist might be impressed with the fact that a mere set of universal hypotheses cannot by themselves be testable. And since one might insist that it is the hallmark of a theory to be testable, then one would have to express all theories in testable form, which would require a theory to be combined with some initial condition or other.

I would regard this as a cumbersome and unnecessary procedure. When we speak of a theory as being testable, what we always implicitly mean is that we know of initial conditions which if combined with the theory would lead to a specifiable test; we never mean that from the theory alone, without any initial condition, a generalisation or an observation can be derived. Moreover the ordinary intuitive use of the term is simpler in the following context: when a statement is put forward as being a candidate for being a scientific theory, the question before us is, can we or can we not find some initial condition that can be successfully combined with it to render it testable. It would be clumsy to adopt the alternative usage, which would be the following: does this candidate for being a theory contain within itself any initial condition enabling a generalisation or observation to be deduced? There is even a more pungent point to be made in this connection, and that is that in striking cases a deduction cannot be affected from a theory, *even* with appropriate initial conditions, *unless* an *approximation* is introduced. In this situation we certainly regard a theory as testable; but the usage under discussion would require us to use the term theory in a torturous and unrealistic way, such that it consisted of universal hypotheses plus an infinite set of initial conditions and an infinite set of unspecified approximations.

Hence I will adhere to the intuitive and I believe well-nigh universal usage according to which a theory consists of the universal hypotheses alone, and excludes all initial or boundary conditions that might be put with it. The most important reason for clarity in this connection is over the matter of approximations, the significance of which in scientific and metascientific works is often underrated. Otherwise expressed, I adhere to the usage according to which a theory is a universal statement which does *not specify* its own tests; and this usage does not even seem to render a theory empty of content (on a more or less realistic interpretation of theories); for a theory would continue to describe or at least claim to describe happenings and things in the world. Thus Newton's theory of gravitation tells us that between two stars there exists a force of attraction which has a certain specifiable quantity. We do not know whether or not this is true or even testable without bringing in initial conditions; but this is what the theory tells us, and so the theory even without any initial condition added is not empty.

Is a single theory testable?

It is not entirely clear what metascientific view of this matter has come to be most influential since Popper first published his *Logik der Forschung*. One idea has been that the test of a theory is a test of a *single* theory. Then there has arisen the idea that the test of a theory is not just a test of a single theory but a test that discriminates between one theory and another. This is very often true. Thus when the Eddington expedition in 1919 was undertaken to make observations of the sun in total eclipse, this test was devised to discriminate between Newton's and Einstein's theories. The situation is often less clear. Nonetheless when we think we are testing a single theory we may find, on closer examination, that we are in fact contrasting it with some other commonsense theory — perhaps not even an articulated one, still less a theory known to science — and this is where it is important that commonsense views should be seen, from a metascientific point of view, to be on a par with scientific theories, however rough and ready or however naive or however unqualified they may be. Thus at some point in our scientific history, the idea that a ball thrown in the air returns because of a pull by the earth might have seemed to be the only theory in existence because of overlooking the idea (to the scientists a mere anthropomorphic one) that after a while the ball had to return to its natural home. And indeed the notion of a 'natural place' was commonly used in physics in days that might now be regarded as pre-scientific but which were in fact quite sophisticated from the scientific point of view. Thus we may go quite a long way in the direction of maintaining that a theory is always tested in opposition to another theory, provided it allowed that this contrasting theory may be a very simplistic or out-of-date or commonsense theory.

Nonetheless the question arises whether this is so in principle. As an example it might well seem that the test for the Newtonian theory of planetary orbits was entirely new, and did not discriminate between two theories. This would be in part a mistake, because Newton's theory was propounded in opposition to or in contrast with the Cartesian physics, according to which celestial bodies were pushed through space in vortices; for the Newtonian theory of gravitation explained the elliptic orbits of the planets. Thus the Cartesian physics provided an explanation of motion, that is, of the fact that heavenly bodies move. But it did not provide an explanation of the elliptic orbits of the planets, and for this the Newtonian theory does appear to have been unique. Put this way, it seems perfectly reasonable to suppose that a theory may be put forward to explain some phenomenon, a test may be devised, and the sole question is very simply, does the phenomenon satisfy this test or does it not, irrespective altogether of whether there is another possible explanation that might do the same work.

Scientific and commonsense theories

A brief word should be added on the relationship between the scientific term and the term 'theory' as used in ordinary life. For there is a widespread belief among purist philosophers that there is little or no connection between the two, and among many there is a wish to repudiate the application of the term 'theory' in matters of ordinary life. It is desirable to recognise that although the term 'theory' may not be commonly used in everyday life, it could be appropriately used and in the same sense as in science. For instance, when we speak of the attitude of the government, we are forming a theory in the same sense insofar as we are referring to something beyond matters of present or past observation and making a corrigible statement, for all such statements share with the use of the term 'theory' the attribute of being fallible. We may find evidence 'beyond all reasonable doubt', but we cannot in principle find absolute evidence affording proof beyond the logical possibility of refutation.

It would even be of value if the man in the street and purist philosophers were to realise that huge numbers of ordinary statements are theories in this sense; it would help to reinforce awareness of the falliblistic nature of such statements and the need for enquiring into the circumstances in which they might be false.

It should be noted that in ordinary life 'theory' can refer to *particular* statements, but it is usually clear from the context whether a universal or particular theory is meant. It should, parallelwise, be noted that the same holds in science. Thus we may speak of the theory that the moon is volcanic.

Observations as theories

Popper has rightly stressed that observation-statements are theoretical. By this is meant that the terms all have reference beyond the immediate present beyond immediate observation. They therefore make a claim beyond what can be verified in the immediate present. Therefore they may turn out to be false: that is to say, they are theoretical.

This is a highly important consideration in the metascience of testing and also in the context where it is relevant to bear in mind that theories are not absolutely falsifiable. But it may well be put to extreme use, and it is important to realise that even though an observation-statement is a theory — perhaps 'theory-dependent' would be better — it is nonetheless a theory on a very different level from a generalisation, and still more from a theory in the main sense discussed above. It will still remain that the metascientific examination of very general statements and of particular observation-statements will be different in certain contexts. The testing operation will be different for the two types, for example, and the role they play in testing other statements will be different in the two cases. So long as this is not forgotten no confusion need arise, through the realisation that observation-statements are theoretical in the broad sense in question. And indeed we may continue to use the convenient shorthand of speaking of observations and their use in testing without reference to their being theoretical, and without having to resort to some more cumbersome phrase such as the testing by a 'theoretical observation-statement'; for this would be irrelevant in most contexts, and unintelligible outside the ranks of those who are familiar with the theoretical nature of observation-statements.

Non-instantiative concepts

This subject is better known nowadays by the heading 'theoretical terms'. There is, I think, one confusion and one problem to deal with.

The problem is created by the (sometimes unvoiced) attitude that 'theoretical terms' are not 'genuine'. It arises solely because of the empiricist-reductionist approach to science, which asserts that such terms are not really meaningful unless we can express them by means of observational (instantiative) terms — in other words, reduce them to experience. This procedure if consistently carried on would in effect do away with the existence of theoretical terms and thus seemingly we have no subject to deal with.

The confusion is to identify the notion that I introduced (Wisdom, 1952) with the notion introduced by Campbell (1921) which has given rise to the description 'Campbellian hypotheses'. The idea that I was concerned with had to do with unobservables, that is to say, things beyond what figure in generalisations. Generalisations give a regular connection between two entities of which instances exist. For example, all gases expand when heated relates to instances of gases, instances of expanding, and instances of heating. Thus all the terms are what I called 'instantial'. By contrast, terms such as Newtonian gravitational attraction, I called *'non-instantial'*. H.H. Price later introduced a refinement, using the phrase *'non-instantiative'* (Price, 1953) which I now follow.

A non-instantiative hypothesis, then, is one that contains at least one non-instantiative term. Not all the terms need be non-instantiative. The general point that impressed itself upon me was that non-instantiative hypotheses and non-instantiative theories refer to what is beyond what is observable or has observable instances, and lie at the root of high-level scientific explanation. At the time I could find nowhere in the literature such an idea isolated, but it must have been intuitively recognised and also it may even have been in the air, for under a year later also in 1952, J.H. Woodger (1952) produced a beautiful treatment — quite independently of course — of this idea which he developed in one Part of his *Biology and Language* illustrated by Harvey's theory of the circulation of the blood. Woodger's procedure was to call ordinary observation-statements of particulars 'zero-level' hypotheses; generalisations which contain only observation-type terms he called 'level-I' hypotheses. Explanatory statements with a non-observation type of term he called 'level-II' hypotheses — and of course there was no limit, though there would be very few examples for level-III or level-IV hypotheses. Newtonian mechanics, for example, would be level-II, the Dalton theory of the atom would be also, so would the kinetic theory of gases, and so would the theory of the gene — there would be perhaps a dozen such high-level explanatory non-instantiative theories in existence.

It might be considered that the name I use is slightly inappropriate, for the non-instantiative term atom, for example, might be held to denote instances of atoms. However this is a hypothetical claim made by the atomic theory; we do not have instances of atoms in our experience. The reference of instantiation is, of course, overtly perceptual; and this is natural enough. After all, when doing science, *our objective is to widen our experience from a knowledge of things that we know of by experience to a knowledge of things beyond our experience*. The point could well be made, then, that the description 'non-instantiative' might have been better if it had been replaced by an overtly non-perceptual term. This may well be so but there is one point in favour of 'non-instantiative', which is that it underlines the conjectural nature of our knowledge of such entities.

Intuitively any philosopher of science can readily see the difference between instantiative and non-instantiative terms as thus briefly described. Not surprisingly, however, difficulties arise if one probes more deeply, for we find that there are cases where we might well be in doubt about which box to put them in. One sort of difficulty arises over attributes of other persons' personality or behaviour. Is anger, for instance, instantiative or not? There is no doubt that in psychological contexts it is most useful to regard statements about people's anger and so on as observation-statements. This may, however, only show that 'observation' is much more elastic and has much more content than it is usually given credit for by purist philosophers. Let us consider one or two examples from physical contexts. Is velocity instantial or not? I would unhesitatingly say that it has instances and we come across them every day. Alas, one gets into difficulty on realising that velocity is a concept constructed out of unobservables, namely, the limit of a vanishing distance divided by a vanishing time-interval. This leads me to interpret terms that are explicable by means of instantiative terms as being themselves instantiative. In this case, distance, time-interval and limit are all instantiative. By contrast, gravitational attraction cannot be so explained. One of the bewildering aspects of physics is that the concept with which celestial mechanics begins is nowhere explained — neither at the beginning of the subject nor in working it through nor at the end. What tends to happen is that the term is mentioned, and it is explained that of course it stands for a vector, that is, it has point of application, magnitude, and direction (thereby the same as a host of other concepts). A self-conscious teacher may point out that it is not at all like force as we experience it when we push a pram or pull a chain, after which he is likely to say that the notion will become clear by using it, and he will slide as quickly as possible into the mathematical derivations from the non-instantiative hypothesis embodying it (he never returns to the subject, however). Some authors over-strongly dominated by pragmatism, logical positivism, operationalism, phenomenalism, or what not, try to squeeze a definition of such terms out of their context, and in fact try to reduce these terms to the other terms of the hypothesis embodying it. It suffices, I think, to remark that the whole point of introducing the term is just because the other terms are insufficient to do the work required, so the new term has to be independent. I take it, then, that non-instantiative terms are explicable in no way, and one has to put up with using them and getting used to handling an unknowable in its context, while instantiative concepts are explicable in terms of other instantiative concepts which have instances plain to see or instances easily found.

Whatever the true nature of the distinguishing characteristic, the discrimination between these two types of terms is now common currency and generally known as a question of 'theoretical terms'. But those who have spoken of Campbellian hypotheses have, I think, confused the theoretical terms in the sense discussed with what Campbell himself seems to have been concerned with. In his otherwise excellent account of the nature of scientific theory, Campbell included an idea which would be regarded by most metascientists as irrelevant: he held that high-level explanatory theory must include ideas that are *familiar*, while Popper, for example, insists that in explanation we are explaining the known by the unknown: that is to say, the high-level explanatory theory must involve things that are not familiar, and this really is the whole point of theoretical terms. Did Campbell then make a rather obvious mistake? I would say not: it was not that he was mistaken but that he was thinking of something quite different. The context that he was concerned with may best be explained by one of his own examples, namely molecules. He pointed out that these are not observable. However, instead of placing the emphasis on the fact that these are non-instantiative in the sense described above, or are theoretical terms, though he surely recognised this, his focus of interest was that they refer to ingredients of science whose existence is not testable: for, as he points out, it is possible to accept the results of molecular theory and yet to deny that there are actually molecules. It could hardly be clearer that Campbell's concern was with what I have described elsewhere (Chapter 15) as the embedded ontology of a theory, that is to say, something over and above its empirical context. Indeed his writing in this context shows that he was preoccupied with the metaphysical underlay of science. Campbell's idea was that such ontological features of the scientific theory had to be familiar in the sense of being analogous to ideas of ordinary movable objects. This feature of his theory carries but little cogency, however, because many theories involve non-instantiative terms with no analogical connection with anything observable. In fact, the more abstract and the more powerful the theory the more the analogy is whittled away, as for instance in the concept of a wave in wave mechanics or force in gravitational theory; and so far from this implying that the more abstract theory is less good, it generally means that the theory is more powerful. However, the point at issue here is that a Campbellian hypothesis should refer to the ontological features of the theory and to his particular view of what constitutes them. The notion that I put forward under non-instantiative terms, however, concerns simply the notion of entities beyond our experience leaving it open *in this context* whether such entities actually exist or not. So much for the confusion involved.

The problem associated with non-instantiative terms arises solely because of the empiricist-reductionist approach that such terms are not really meaningful unless we can express them by means of instantiative terms — reduce them to experience. Bold attempts have been made to this end. One was made long ago by Ramsey (1931). One that has been accorded particular attention was afforded by a logical theorem discovered by Craik (1953, simplified version 1956). Hempel has given an account of Ramsey (Hempel, 1965, pp.215f) and of Craik (Hempel, 1965, pp.210−15) together with a full discussion of the problem (Hempel, 1965, pp.185−226), which he calls the theoretician's dilemma. The dilemma is as follows. Given that instantiative generalisations provide connections between observations, let us suppose that from a certain non-instantiative theory T we can deduce these same generalisations, all of them and no others. Then T is replaceable by the generalisations and is therefore dispensable; that is, the deductive systematisation achievable, for example, T, is exactly the same as that achievable by the generalisations; hence if T serves its purpose it is unnecessary. This is Hempel's way of bringing out the kernel of Craik's theorem. However, he also brings out what Craik himself recognised, that, broad though the scope of the theorem is, there is a price to pay for eliminating T in favour of the generalisations: namely that these have to be qualified by postulates consisting in *all* the distinct (that is, non-equivalent) observational generalisations of T (not merely a subset) and, worse than that, there would be an infinite number of such postulates. Hempel concludes that the manoeuvre is not playable. Craik's theorem is a proud achievement to the reductionist end, not worth attempting if the end is not acceptable, not worth achieving at the cost of conditions that cannot be fulfilled or which are even more difficult to swallow than the irreducibility of non-instantiative terms — but the theorem might be interpreted as virtually a reductio ad absurdum of the policy of liquidating non-instantiative terms.

From a different viewpoint there is no problem at all. On the contrary, non-instantiative hypotheses are what we *seek* in science, because of the putative attempt they make to give us knowledge beyond our experience. The empirical context of a non-instantiative theory has the function of *explaining* phenomena, generalisations, observation, problems; the embedded ontology of a non-instantiative theory, to borrow from Hattiangadi, has the function of *describing* the structure of the world (Hattiangadi, 1978, 1979).

Are non-instantiative theories needed for scientific discovery?

Consider a form of natural science containing only instantiative terms. That is to say, it would consist of a host of generalisations.

It might be possible to assemble some or all of them into a wider generalisation involving only instantiative terms. An elementary example, which I (Wisdom, 1952, pp.23—5) have drawn from physics, will show what is meant. One gas law is that the volume, v, of a given mass of gas is proportional to the absolute temperature, T, provided the pressure, p, is kept constant. Another is that p is proportional to T, provided v is constant. Now these two laws can be unified, allowing p and v to vary, by the single more complex law that the product, pv, is proportional to T. And this law will do all the work of the other two. Here, granting that the two initial laws are instantiative, so is the more complex combination.

Such a consummation, almost never articulated, would seem to be the ideal in the minds of many natural scientists and metascientists. A grand generalisation, yielding all others as particular cases applicable in special conditions, would be highly explanatory and a great achievement.

Two questions suggest themselves. Would anything be missing which we should like to see included in our scientific account of nature? And why do we not possess a pattern of wider and wider generalisations building up to a grand generalisation in natural science?

I would concede that we might well be satisfied with a grand generalisation. Such a concession would not amount to much, however, if it were impossible to achieve such a goal. And — the second of the above questions — it is possible to see why we find nothing approaching such a pattern in natural science. For in natural science we have laws, for example, about magnetic attraction and laws about electric circuits, or laws about planetary motion and laws about the motion of comets. But there is no *specifiable* way we can find a larger generalisation to encompass magnetic attraction and electric circuits, or to encompass planetary motions and comets. Strangely enough, *easier* than the task of finding such larger generalisations is the extraordinary procedure to be found in natural science of inventing an explanatory theory with non-instantiative terms, such as Maxwell's theory (electro-magnetic equations) or Newton's gravitational theory. The curious outcome of such theories is that *they yield* the larger generalisations that we are unable to find otherwise. Therein lies the power and beauty of theoretical explanation with non-instantiative terms. Moreover, we can see that they *could not* yield such generalisations, containing, as they must, purely a mixture of instantiative terms unrelated, in our experience, to purely instantiative means; for if they could, this would be equivalent to our finding the larger generalisations directly in the way we find any smaller generalisations.

In short grand generalisations would do for certain purposes if we could find them. (Computers *might* turn up some, but probably only to a *very* limited extent.) A grand generalisation, not a grandiose one for all physics, but one yielding a dozen or a score of simple ones, would serve for practical calculations; they would tell us about the operations of nature. Thus a triple-grand embracing Kepler's laws would tell us about sputniks, meteors, moons, and planets. Non-instantiative theories will do all this, too. They will, in the second place, yield curious generalisations such as that heat has minute weight (see, for example, Popper, 1976), which would otherwise be hard to take seriously. But, thirdly, their most striking property — bad in the eyes of some — is that, even if tentatively and untestably, they describe the structure or nature of our cosmological furniture: they do what the mind of man has always sought, they tell us what the world is like. Still more shortly, they are pragmatically invaluable (so far as one can tell they will never be superseded by the only known alternative method), and philosophically answer some of our deepest aspirations.

Summary and conclusion

If the main problem about the notion of 'theory' arises because of (a hangover of) a positivistic attitude that wishes to achieve reductions everywhere — that is, if this is a man-made problem that has produced quite a good livelihood from one another's washing — there are nonetheless some real and highly important matters to do with routine clarity and one genuine problem.

1 Because of a fuzziness in common scientific use, it is desirable to try to fix sharply a difference of reference between 'theory' and 'generalisation' or 'law'. All natural science practitioners intuitively know the difference; but outside their ranks we quickly reach a tacit attitude that 'generalisation' or 'law' includes the lot, that is general statements about what goes on in the world, which we can perceive. The confusion here has made it possible for philosophers to prolong the life of the belief in induction; whereas its natural plausibility would otherwise have been totally undermined. In the social sciences the confusion strangles fundamental research theories at birth.

2 Failure to distinguish 'theory' obscures the basic process of testing scientific knowledge.

3 It aborts the possibility of discovering by its means new generalisations not otherwise known.

4   And aborts the possibility of imposing on old generalisations moderate new corrections.
5   Without 'theory' the avenue is blocked off to the enormously important role of *approximation* in science.
6   The issue never gets off the ground about the function of scientific theory in relation to the world — what it should tell us about the world.
7   Then comes from theory, what cannot come from generalisations or laws, the *possibility of correcting observations*.
8   And a problem opens up about the theoretical aspect of observations.
9   It is here claimed that the peculiar nature and power of theory springs from the ingredient of 'non-instantiative' terms.
10 It is claimed that so far from being a blemish, they are the crowning glory of science. And perhaps Hattiangadi has added a mighty cairn to the mountain of science by showing that their function is to *describe* the structure of the hidden world (Hattiangadi, 1978, 1979).
11 Hattiangadi would thus be showing in a new way the key role of 'truth' for science.
12 I have added a new argument to show that non-instantiative terms and theories are indispensable to science because they can lead to new generalisations almost certainly not discoverable otherwise.

References

Campbell, N.R. (1921), *What is Science?*, New York, Ch.5.
Craik, William (1953), 'On Axiomatizability within a System', *Journal of Symbolic Logic, 18*, pp.30–2.
Craik, William (1956), 'Replacement of Auxiliary Expressions', *Philosophical Review, 65*, pp.38–55.
Hattiangadi, J.N. (1978), 'The Structure of Problems, Pt.I', *Philosophy of the Social Sciences, 8*, pp.345–65.
Hattiangadi, J.N. (1979), 'The Structure of Problems, Pt.II', *Philosophy of the Social Sciences, 9*, pp.49–76.
Hempel, C.G. (1965), *Aspects of Scientific Explanation*, Free Press, New York.
Kaplan, Abraham (1964), *The Conduct of Inquiry*, Chandler, San Francisco, Ch.2, Sect.7.
Popper, K.R. (1976), 'The myth of the framework', *The abdication of philosophy: philosophy and the public good*, Open Court, Chicago, 47n.
Price, H.H. (1953), *Thinking and Experience*, London.

Ramsey, F.P. (1931), *The Foundations of Mathematics*, 'Theories', Routledge & Kegan Paul, London, pp.212—36.
Wisdom, J.O. (1952), *Foundations of Inference in Natural Science*, Methuen, London, Ch.3.
Woodger, J.H. (1952), *Biology and Language*, Cambridge, pp.32f.

PART III: REALITY, WELTANSCHAUUNGEN AND DISCOVERY

14 Conventionalism, truth, and cosmological furniture

The problem to be discussed here concerns ontology so far as it may *not* be formed by scientific theory. In brief terms, the problem arises in the following way. On the one hand, the world surely consists of whatever is there, irrespective of whether human beings are around or not, and irrespective especially of whether human beings have constructed any scientific theories depicting the nature of the world; on the other hand, scientific theories are subject to the limitation that we can never verify them or prove them to be true, so that the ontology prescribed by scientific theory is not firmly established but is only what is attributed to the world by a fallible scientific theory at any given time. The problem could also be put in terms of 'conceptual networks' or in terms of language-systems. I shall develop various aspects of this clash and only gradually come to a sharper formulation of it; thus it is not restricted to the philosophy of science but is a problem for philosophy in general. But I think its sharpest form comes from the fallibility of science, aided by the consequential power of conventionalism. The problem will become more specific later.

Philosophy of science: conventionalism

There are various interpretations of the nature of science. I have concluded elsewhere, however (Wisdom, 1970, 1971), that of these conventionalism is the only serious claimant, apart, that is, from Popper's (for not even instrumentalism, wide following though it has, is viable at all). I concluded also that conventionalism was hardly viable, that is, that it is up against the most formidable difficulties; nonetheless, it may not be utterly beyond the bounds of being rescued; and it might have to be rescued in view of a serious difficulty in the realist alternative that goes with Popper's approach, as I will bring out below. Popper himself has also treated conventionalism with considerable respect and has noted that it differs only to a very small extent from his own view — though the difference that remains is a vital one. Thus, like Popper's view, conventionalism looks on science as starting with a problem, forming hypotheses or theories, handling them within the hypothetico-deductive framework, making deductions therefrom, making observational predictions, testing these, revising the hypotheses or theories in the light of the tests, noting that absolute verifiability or proof is unattainable in principle and noting also (as is often forgotten about Popper's theory) that absolute refutation is impossible in principle. With this extensive overlap between conventionalism and Popper's view wherein lies the difference? The difference is that on Popper's view a hypothesis or theory tells us something about the world; for conventionalism on the other hand, it does not. It provides an 'as if' account.

One of the underlying sources of the problem lies in the difficulty of refuting a theory (and of course of verifying it). Duhem's chief contribution was probably his emphasis on the fact that when a test turns out to be false and constitutes a refutation, it refutes not a single hypothesis or theory, but the entire theoretical structure of hypotheses, initial conditions, and so on, from which the test observation is deduced; so that there is no means of knowing which ingredient has been falsified.[1] This highly significant conclusion was, of course, taken over by Popper (1959) although absorbed in his realist interpretation of the nature of science. But Duhem in effect took it over from Poincaré (1952, originally 1902) who, though he may not have articulated it quite so overtly as Duhem, did make it virtually explicit; the realistic difficulties of refutation certainly come over strongly in Poincaré. We have now to make the step from irrefutability to conventionalism.

To illustrate the conventionalist approach more concretely, if in geometry we alter the conventional meaning of the term 'straight line', we obtain a new system which is just as valid as the initial one. Thus, Riemannian geometry is just as valid, as a system, as is Euclidean geometry, but in it the term 'straight line' assumes the new meaning of geodesic, which is the shortest distance between two points, though it is not straight in the ordinary Euclidean sense (thus a great circle on a globe is geodesic). The central idea here is that one theory may be exchanged for another one by altering the conventions to do with the meaning of basic terms. And here we have the central idea of 'truth as a convention', not as recording something about the world.[2] Expressed conveniently though metaphorically, a theory can be regarded as a pattern, drawn not because it reflects reality but because it brings order of some sort into our observations; and this order or pattern is conventional, because, if it does not produce adequate order, we can modify the pattern to make it fit together. In effect, theoretical terms on this view have no denotation. (This has been neatly formulated by Agassi, 1966, pp.4–5.) The relationship between framing conventions and ordering our observations is that order is the basic objective subserved by manipulating the conventions suitably.

Such a procedure interprets smoothly a peculiarity of science: namely, that a theory is basically irrefutable, that is, irrefutable by observation; for even if a theory does not square with its predictions and if for the time being no alternative convention is available that would make it square with them, nonetheless there is always the possibility that a new convention might be thought of that would save the theory. But if irrefutable, neither is a theory conclusively verifiable: there always exists the possibility of putting together an alternative theory that would equally well explain the phenomena covered by the given theory and lead to the same predictions as the first one; but since no theory is refutable, the alternative theory could not be refuted, and, so long as this is the case, the given theory could not be regarded as conclusively verified. Thus acceptance of a theory would be conventional because it would be an arbitrary choice of one out of a set of possible theories that would do the same work.

Thus, a theory cannot be accepted or rejected on the grounds of being true or false in the light of evidence (for the evidence is not conclusive even though it is all we have); it can be maintained or retained simply by adjustment of the meaning of a key term, that is, by changing the conventions for the use of a term.

Putting these facets together we have the following: First conventionalism starts intellectually from the irrefutability of theories in an absolute sense, since a discordant observation refutes a system as a whole and not the theory alone. Second, dwelling on this we come to its consequences: that alternatives are also irrefutable, and that no theory is verifiable; in short the truth-value of theories cannot be absolutely settled by evidence. However, though refutation is not absolute, evidence is sometimes interpreted as being against a theory. Third, all this fits agreeably into a setting in which it is recognised that the meanings of terms can be adjusted by convention, and adjusted in such a way as to preserve the theory against the discordant evidence, that is, to preserve its 'truth' (the realistic difficulties of refutation encountered by Poincaré fell on a suitable soil, the impact upon him of alternative geometries).

Irrefutability alone, without making the third feature explicit, is not sufficient to constitute conventionalism. All that it asserts is that, since we cannot possess conclusive evidence, therefore we do not have *absolute* knowledge, and therefore we do not have *knowledge* — only 'as if' theories, that is, our observations cohere *as if* our theories record something about the world. Such a conclusion may give the erroneous impression of being very cautious, since, in the absence of any basis for the truth of a theory, there can be no such thing as a scientific description of the world even in principle. In fact, this sweeping negative conclusion is the reverse of cautious. All that is warranted is that we can never tell whether or not our theories describe the world, and this is Popper's line: a scientific theory may very well be absolutely true but we could never know it. Now if this modest agnosticism were all that was being claimed by conventionalism, then conventionalism would hardly be an outstanding interpretation of the nature of science. Let us therefore turn to the other main ground for the doctrine. It is true that Poincaré was overwhelmingly impressed by the difficulty of refutation, and quite rightly so, but I would suggest that this only *raised* for Poincaré the problem of the truth of scientific theories. Poincaré did not erect conventionalism on this basis alone; and in support of this last point it must be borne in mind that it is impossible to think of Poincaré except against a background of the great modern developments of geometry and algebra, not forgetting that he was himself a contributor to this geometry. The decisive factor would seem to be the consistency or, if we will, the coherence of these modern geometries, which are just as consistent as Euclidean geometry, with *no way of choosing* between them — it is this that provides the raison d'être of conventionalism.

In short, the impossibility of pinning falsity on a theory, because falsity may lie anywhere in the whole theoretical system containing the theory, paves the way to the real basis of conventionalism; namely that although a theory is irrefutable in principle, it is *threatened*; but the threat can always be removed and the theory saved by altering the conventions.

In this background, we may notice the confluence of *irrefutability*, *equivalidity of theories*, and *meaning by convention*. Irrefutability lies at the root of equivalidity, and equivalidity leans on the possibility of changing the conventions — or so, at least, it would look to anyone immersed in four-dimensional geometry, for the different systems of geometry may be interpreted as a difference of convention about the meaning of the term 'straight line'. If we extend this notion of conventionalism to the special theory of relativity, we get the result not that the three-dimensional metric of Euclidean geometry is *false* and that Einstein's four-dimensional metric is *true* but that the three-dimensional *convention* yielding the metric has to be *replaced* by the four-dimensional *convention*. In short, it is not that there are different basic forms of conventionalism, but that it contains a tiered structure of irrefutability, equivalidity, and conventional meaning, each important in some contexts, the last being metascientifically basic.

There would seem, however, to be a deep-seated oversight in the idea that we can preserve a theory by altering its meaning by convention. The device of convention-replacement will, of course, enable us to retain certain parts of a theory. Thus 'A straight line is the shortest distance between two points' remains true in four-dimensional geometry as it is in Euclidean geometry. That is to say, certain relatively *unspecific* propositions may be retained. But if we look to the more *specific* propositions that come under these, which are after all characteristic of the various theories, then these are simply not retained and are not equivalent at all in alternative theories, not even by convention. (Poincaré showed his concern about this in his discussion of the principle of conservation of energy.) Thus, the Galilean metric as measured in a moving frame is not the same as the Einstein metric, and the scientific position is first that both cannot be true and second that the evidence is against the Galilean metric.

Since conventionalism, even in the more refined Poincaré version, has something of the air of being a contortion, the question arises what is the underlying leitmotif for such an approach. The conventionalist ploy has worked beautifully in some areas. There are many such processes that have worked very well in mathematics and logic, but which do not seem applicable in empirical science. Be that as it may, it seems to me that this was the underlying factor that provided Poincaré with a general approach to knowledge. But we are still faced with the difficulty of understanding why this made such an impact, even granting that he was imbued with the difficulty of getting decisive evidence for or against anything in science. The reason is, I think, that this position rests upon an intuitive sense of difficulty about a realist interpretation. So, if one could not get truth, then the next best thing is a satisfactory convention. That was the leitmotif. The problem here has come out into the open, since conventionalism has reappeared in a virulent form, to the effect that rational comparison of theories is impossible if they contain 'incommensurable' terms.

The strongest support for conventionalism of recent times has come from the 'incommensurability-thesis' introduced by Kuhn (1962) and by Feyerabend (1962), which was later given a somewhat different form by Lakatos (1968, 1970*a*, 1970*b*); this will be discussed in an appendix to the chapter.

The coherence theory of truth, conceptual networks, and other influences

It is worth remarking briefly that conventionalism is the scientific equivalent of the philosophical doctrine known as the coherence theory of truth. According to this, truth consists in a body of mutually consistent or internally coherent system of propositions such that no outside criterion is needed to guarantee the truth of the system (though a further internal criterion may be required, such as 'pervasiveness of the system'). The reason why the coherence theory is not as a rule regarded as cogent lies in the fact that even a complex system could be fully coherent and yet false in the following sense: internal consistency alone would merely give the system an ivory-tower existence; so at some point, even if it is at only one point, the system would have to connect with the world of reality; that is, a coherent system would not have to touch reality at many points, but at least one proposition would have to be true of the world — in short, the coherence theory must depend at some point on the correspondence theory.

On the other hand, various movements of contemporary thought serve to support the coherence theory, even though not explicitly, for example, cultural relativism (overtly interpreted as conventionalism by Jarvie, 1964), schizophrenic delusions, interpreted as true of the patient's own world, sociological epoch-relativism, the philosophical notion of a conceptual network, and the notion of a 'construction' in connection with historical fact, criticised by Nowell-Smith (1971b).

It is assumed here that all these attempts at support can be defeated; and that conventionalism would have no force were it not for a difficulty in the alternative.[3]

Ontological realism

The realist view is that scientific theories claim, either successfully or unsuccessfully, to describe the world. Popper (1959) and Agassi (1966, pp.21ff) both claim that science has its foundations in realism. If there were no difficulties about this, there would be no compelling reason for trying to give conventionalism a run for its money.

Associated with realism is the correspondence theory of truth in some form or other. The broad intuitive sense of this is simple enough; namely, a theory is true if it describes, or records something about, the ingredients of the world. But how is this possible? For the terms of the theory reflect conceptions of ours which may spread their net far more widely than the part of the world they are supposed to depict. Turn from the case of scientific theory to the ordinary case of snow. When we talk about the material, snow, we do so by means of the word 'snow', but in delimiting the material snow by means of the word 'snow', we are imposing upon the world a hypothesis that there is a stuff, snow, and other stuff, non-snow, from which it is distinguished, and the hypothesis that snow has such and such recognisable characteristic beyond our present observation. However, we do not know whether this is the way the world is. All we know is that these are our ideas about it, or, if we prefer, that these are our experiences of it or that these are hypothetical; we presuppose the existence of experiences, conceptions, or terms which may be much richer potentially than the materials of the world themselves. All this is commonly expressed by the contention that concepts such as snow and white are 'dispositional'.

Here, however, I wish to make this point in a different way. Assuming that all observations are theory-laden (leaving aside the question of how we might know this), then what becomes of the realist theory of truth? An observaton-statement is true, if and only if the subject-matter of the observation is present; but if this subject-matter is theory-infused, what is the world like apart from our theories?[4] It has been put by Quine (1969) in a slightly different, and possibly more fundamental, form: 'it makes no sense to say what the objects of a theory are, beyond saying how to interpret or reinterpret that theory in another.' Quine describes his view as 'relativism'.

Gestalt psychology, too, would apparently support the same sceptical tendency. Again, if we adopt a phenomenological approach, the dichotomy is altogether repudiated between experience (expressed in a statement) and a reality to which it corresponds.

What is happening here is that the real world is beginning to slip through our fingers so that the haunting suspicion arises that there is nothing actually present for the theory to correspond to.

If we are linguistically minded we can express the same point within the terms of that approach; for an observation would be described by our language, and we can never be sure that our language has got sufficiently close to the observation. Moreover, in terms of the idiom widely used in philosophical analysis, 'seeing as', we may note, for example, that from an aeroplane we see clouds 'as being' a field of snow, and on the ground we see a huge pile of white flecks 'as being' snow.

What are the bare bones of the reality? If reality is partially constituted by theories (or by the nature of our language), what is the 'solid core' of reality that is not so constituted? Is there no *cosmological furniture apart from our theories* (apart from our language-systems, apart from our phenomenological experience)? Or did Galileo give vent to what excites scientists about science (his reference was to mathematics) in regarding it 'as opening to us the very heart of nature'?[5]

The implication for Tarski's definition of 'truth'

It is of interest, though it is an incidental matter,[6] to view this difficulty in terms of Tarski's (1956) celebrated definition of 'truth', which is widely believed to have successfully overcome the difficulties of the correspondence theory. I shall express it in the following form:

'That snow is white' is true, if and only if, that snow is white.

The principle involved enables us to assert

'$F=M_1M_2/r^2$' is true, if and only if, $F=M_1M_2/r^2$

Wherein, then, lies the problem?

The problem disclosed above arises out of the right-hand side. The right-hand side is described in terms of the theory, as if the terms of the theory recorded in some way directly the ingredients of the world. How does this affect Tarski's definition? The terms of the right-hand side cast their net more widely than the actual ingredients of the world they are supposed to record, all the way from abstract theoretical terms to observational terms. Assuming then that observations are theory-laden, we now have to reconstitute Tarski's definition in the following way:

> 'That snow is white' is true, if and only if, the theory-laden-observation-snow is characterised by the theory-laden-observation-white.

And this would seem to render Tarski's definition useless, or at least no help to the correspondence theory of truth. Can it be saved?

The explicit problem

We are now in a position to grasp the problem more precisely. So long as the realist or correspondence theory of truth is taken for granted, there is nothing compelling to be said for conventionalism despite certain points that can be made in its favour. But there is some difficulty over the distinction presupposed between the world on the one hand and our theories or statements on the other that correspond to it. For this distinction is at risk (a) because we cannot ascertain the nature of the world apart from our theories or statements about it and (b) because of the thesis or assumption that observations are theory-laden; and these difficulties might seem to be so great as to lead us back to some form of conventionalism.[7]

The relation between thought and reality

Whether we can conduct our investigation by means of concepts, terms, or words, the relation that such a conceptual element bears to some piece of the physical world has been viewed in a number of very different ways. One is that the conceptual element is an image mirroring a piece of the jig-saw puzzle constituting reality. Another is that the conceptual element is a refined version of the piece, namely, its essence. Another is that the conceptual element has no such relation but is a word such that, when we combine words together in certain orderly ways, we know what further pieces of the jig-saw puzzle to expect. An interesting one was put forward by Wittgenstein (1921) in his *Tractatus*. The model used there was of a fish-net. You could describe a slab of the world by noting what pieces were caught by the mesh of the fish-net, and the description would become more accurate if you reduced the size of the holes. There is obviously something in this, not altogether unlike the idea of certain painting techniques — it is possible to obtain an accurate portrait without drawing any lines by covering the canvas with huge numbers of dots alone.

It would seem that we have here a characteristic clash between structuralism and functionalism: some of the views interpret the conceptual element as representing in some way or other a physical piece and as being *substitutable* for it for purposes of communication; and there is no reason to think that this is wrong so far as it goes, though it might very well be wrong if it is taken to be an exclusive theory. The other view, functionalism, looks on conceptual elements as optional pieces of machinery. This seems to be valid also and wrong only if it is taken to be an exclusive theory. As in other fields, we should here be in a position to admit without more ado that conceptual elements have both a structural and functional job to do.

Neither of them, however, separately or even jointly, allows for the possibility of a piece of the jig-saw puzzle being actually constituted in part by the conceptual element. The notion of the mind as 'pure act' has been outlawed in favour of its being a tabula rasa from the more influential philosophies for so long that philosophical minds have become like the doctrine they have settled for (or at least their successors whom they have taught have incorporated that doctrine so that the doctrine has operated rather like a self-fulfilling prophecy).

I want now to make a modest addition-cum-modification to the classical view of denotation, which perhaps amounts to a referential theory distinct from the correspondence theory. This suggestion may not cover every aspect of the difficulty arising out of the theory-laden nature of our observations, but it seems to take care of a highly important one, holding out hopes for some such mode of solution. It seems to me that a method of splitting up the denotation of a concept can be devised to have a fairly definite core and a rather indefinite boundary. I will distinguish these as a *soft core* and a *hazy boundary* or *fringe*.[8] I distinguish the core as 'soft' to avoid any suggestion of an 'essence'. Consider 'blue', for example. This covers a wide denotation from highly specific shades of dark blue to light blue, green-blue, and so on. However, there is a boundary of a certain width such that in this no-man's-land it is hard to decide whether the colour is blue or green. This may be regarded as a hazy boundary or fringe between a soft core of blue and a soft core of green. There is no reason whatever to think that in rerum natura blue things are separated off sharply from things of other colours. In our conceptualisation, we tend to think of a typical blue or of a few typical blues but forget about the hazy boundary. What I am suggesting here is that the typical blues do form a soft core which is different from the soft core of greens in actual reality, and yet that there is no objective classification in nature. Nature does not classify. Blue things are not separate from green things with a sharp boundary between them, that is, there are no 'essentially' blue things or 'essentially' green things. I have used an example from colours because of its simplicity, but the example will be found unsuitable by some because of the subjectivity and relativity of colours, though this is strictly irrelevant to the point at stake. Another example could be given in terms of 'baldness'. A convenient one to be used below will be the ranges: snow, melting snow, sleet, rain.

(In a similar way, it has struck me that our use of arithmetic in everyday life is often inappropriate. Numbers are sharp instruments. If we say that something is a hundred centimetres, we think of it as being a hundred centimetres, no more, no less. But we do know, of course, that this is not so. The reality is that we measure, say, with a tape measure that can stretch a little bit, and the object we measure may have rather rough ends, so that all we can legitimately end up with is a measure of a hundred centimetres plus or minus .5. This is very clumsy and we should manage much more conveniently if only we had an arithmetic of 'blunt' numbers, such that a name such as a hundred could refer to a length with a hazy boundary. What we need, in order to avoid both discreteness and continuity, is step-functions.)

This general point seems to have some importance in view of legitimate criticism of the metascience of testing given by Körner (1964). The point he makes is that a test-observation predicted from a theory is never exact, and if we take the inexactness seriously we have to admit that the predicted observation is false; in other words, *every* prediction derived from a theory is false, *every* theory is falsified by *every* observation and experiment made as a test. Since Körner's conclusion would tend to support conventionalism, we might enquire whether we can have a theory of observation-statements such that they can, in principle, be true, that is, true in the sense of recording a piece of the world. What I am suggesting is that this can be done by taking more carefully into account the notion of limits of error, which are usually ascertainable in the physical sciences, though usually not in other sciences; and I think this is most easily conceptualised in terms of a soft core and hazy boundary.

Let us apply this notion to the right-hand side of Tarski's definition of truth. In rerum natura, there is no sharp classification separating snow (with hazy boundary) and rain (with hazy boundary); we may have a range from rain to sleet to melting snow to snow, but the soft core of snow is sharply separated from the soft core of rain. And similarly with white.

It would seem that this criterion for an observation suffices for deciding whether or not a prediction from a hypothesis confirms or refutes it. If a hypothesis requires a soft core prediction, then it is refuted not only by a soft core outside this, but also by the hazy boundary or fringe; if on the other hand the hypothesis allows that the prediction shall be satisfied by an observation in the fringe, then it will be refuted only by an observation of a solid core (or of a fringe outside its own fringe).

This is, in effect, a theory of restricted reference. It seems to go some of the way towards supporting the part of the realist view of the world of the correspondence theory of truth: that is to say, in the sense that part of the meaning of what is asserted by an observation-statement can be true of the world, though part of what is asserted at the fringe cannot be taken literally. Otherwise expressed, a statement, to the extent that it is precise, is a precise statement about a soft core but not a precise statement about a fringe.

Here, however, a derivative problem arises.

Is there a sharp distinction between a soft core and a hazy boundary? We know there is in some sense, from examples such as those concerning blue or bald, and the only problem is to provide a satisfactory account of it. The difficulty here is that a sharp boundary cannot be drawn between a soft core and a hazy fringe, from the mere fact that the haze is continuous and fades away rather than reaches a sudden end. I will now suggest a way of handling it.

(Those with an interest in mathematics may note that we can use a device analogous to that used to describe the value of a function as it approaches a limit it cannot reach. Thus the value of the function $1/n$ as n approaches infinity is zero, even though the function can never reach zero no matter how large n becomes. The standard way this has been handled since its introduction by Cauchy is to say that the *difference* between the *actual value* of the function and its *unattainable limit* can be made as small as we please — for which he gave a precise expression by a special procedure.)

Between a soft core and a fringe we can insert a fringe$_1$; and between the core and fringe$_1$ we can insert a fringe$_2$; and so on. My assertion would be that as the successive fringes eroded the central core, they would break off from the original fringe and leave a gap between the later fringes in the sequence and the original fringe. Thus when rain changes to melting snow to sleet to snow, as 'it' reaches the domain of sleet and snow it ceases to be rain, or leaves a gap between itself and rain.

Cosmological furniture: theories and reference

In what relation do theories stand to the world? Let us grant that an observation, that is, a test-observation, being theory-laden, cannot mirror a piece of the jig-saw puzzle composing the world in the way that the classical correspondence theory of truth would have it. Nonetheless, the account just provided for the use of these test-observations is such that they can genuinely test hypotheses and theories. What we have to suppose, therefore, is that hypotheses and theories may 'correspond to' the structure (as well as the functioning) of the jig-saw puzzle. In other words, what we are doing here is to provide a form of the correspondence theory applicable to theoretical statements. This step may be expressed by adapting Tarski's definition so far as theoretical statements are concerned. Instead of writing

'$F=M_1 M_2/r^2$' is true, if and only if, $F=M_1 M_2/r^2$

we write

'$F=M_1 M_2/r^2$' is tentative, if and only if, appropriate soft-core observations (such as the elliptic motions of the planets) actually take place.[9]

A significant corollary from the solution proposed here is that, while the concept of truth applies both to theoretical statements and to observation-statements, the latter regain some of their pristine distinction from the former even though they are theory-laden. Observation-statements are in a certain way theoretical, but this does not justify their being lumped in with theoretical statements to the extent of obliterating the difference between them. After all, human epistemic enquiry begins, as Popper holds, with a problem that arises because an observation (however much infused with certain expectations) conflicts with some expectation. If we were non-sentient animals lacking observations, problems would develop differently; but we are sentient and this imposes a certain structure on the way we enquire and discover. Thus, an observation-statement *is true* if the soft core it refers to, even though theory-laden, is so; a theoretical statement *may be true* (true in the same sense) if a certain observation-statement is true.

There is, no doubt, cogency in the idea that, if all statements are theory-relative/culture-relative/language-relative, we can say nothing specific about nature in itself. We can apparently (at most) say 'something or other is the situation/exists/is happening'. However, we do know a little more than that: 'The situation millions of years ago, however unspecifiable or kaleidoscopic it may have been, even allowing the conventionalist interpretation of scientific theory or of our language, was markedly different from the situation now, taking this also to be unspecifiable or kaleidoscopic.' What I am relying on here is the truism, 'change occurs' — in fact I am using Tarski's definition in the form:

'Change occurs' is true, if and only if, change occurs.

This way of utilising the correspondence theory is different from the one that focuses on a soft core: it applies where there is no soft core at all, and the entire reference is hazy (or cloud-like) — where the hazy boundary or fringe encroaches all over and swamps the soft core, as where a knitted shawl unravels and the fringe absorbs it. I would say that this truism is an invariant for all weltanschauungen[10] (and therefore precludes incommensurability of weltanschauungen). Thus theory-ladenness does not rule out statements about the nature of things.

Being in possession of a theory of reference, we are now in a position to enquire whether it takes care of the problem arising from the theory-laden nature of observation. The soft-core observation of snow embodies a 'dispositional' tendency or the hermeneutics of 'intentionality' or simply is laden with the interpretation that the snow has an underneath and is cold. But these, despite their being humanly infused with ideas, may also be soft core phenomena in rerum natura and compatible with Tarski's definition — the fact of their being theory-laden means only that they are *challengeable*, but *challengeability* does not preclude the possibility of *reference* to a *soft core*.

Where, now, does conventionalism stand? It is not *refuted* by the preceding argument. All that that argument is designed to do is to show that a realist type of theory is *possible*. But given the enormous weaknesses of conventionalism noted elsewhere — and we would not explore conventionalism further were it not for the difficulty of the correspondence theory or the apparent impossibility of the realist thesis — these weaknesses and the *possibility* of a realist theory of reference together suffice for the rejection of conventionalism.

The conclusion is that the correspondence theory holds for specific soft core meaning, for general statements where the meaning is hazy throughout, but not for the hazy boundary of specific soft core statements; that this restricted application suffices to circumvent the difficulties of the correspondence theory, thus undermining the need to try to save conventionalism; and that, despite the theory-relative nature of our ontology (or despite the nature of phenomenological experience), by means of this restriction we can give an account of the furniture of the cosmos.

We could express the problem and the solution in another way thus. Nature does not classify; only our theories classify; are we therefore doomed to total ignorance of what there is in rerum natura; or can we not record some of our cosmological furniture by means of the soft cores denoted by the terms of our theories or at least of our theory-laden observations?

Appendix: The Incommensurability Thesis

Two simple considerations have been given inadequate or no attention concerning the incommensurability thesis introduced by Kuhn (1962) and by Feyerabend (1962) and later given a somewhat different form by Lakatos (1968, 1970a, b).

The Kuhn-Feyerabend thesis is this. Two theories are incommensurable (for example, the Newtonian and Einsteinian theories of motion) if they contain a basic common term whose meaning or use in one theory is incommensurable with its meaning or use in the other, that is, if at least *one basic term* used in both theories has a totally different meaning in each. Incommensurability renders it impossible to compare or contrast, relate, or otherwise discuss the content of the two theories. They differ in relation to the problems they solve or give rise to; but no scientific/rational procedure plays a decisive part, or perhaps even an influential part, in deciding between them. Conventionalism of one sort or another is a natural sequel. (Incommensurability could equally be a consequence of conventionalism; according to Skolimowski (1967), the notion of incommensurability resulting from conventionalism stems from a paper published in 1934 by Ajdukiewicz.)

Simple considerations about the incommensurability thesis might have shown in advance that it must be untenable. Now Hattiangadi (1974) has put forward a short and simple argument to the effect that there are two commensurable theories of the origin of the moon, though neither theory has any connection with the terms of its test. Hattiangadi (personal communication) has, moreover, pointed out that, even if the acceptance of a theory depends on the scientist's assessment of what problems are solved by the alternative theories available, observational tests do exist that discriminate between the exemplars of supposedly incommensurable theories, Newton's and Einstein's (thus the three classical tests discriminate fairly sharply between them, and hence decide scientifically/rationally between them[11]).

Even sharper, however, is the following. The metric or, more simply, the notion of a length is held to have a different meaning in Einsteinian mechanics from Galilean mechanics: in the one, it is a three-termed relation between the distance to be measured, the measuring-rod, and the frame of reference, whereas in the other it is a two-termed relation independent of the reference-frame (supposedly providing different structures for their meanings). And the two concepts of length are held to be incommensurable. It is overlooked that the Galilean concept is *used in relativity mechanics*: it is used when dealing with a length in a frame of reference moving with it!

A most interesting corollary to the thesis may be derived. I will call the form stated above the 'soft thesis'. The corollary, which I will call the 'hard thesis', is that if *one* of the basic terms used in both theories has a totally different meaning in each, then all the basic terms have also. An indirect proof may be given as follows. Let the incommensurable theories be T, T'; suppose, if possible, that there are basic common terms with common meanings $c_1, c_2, \ldots c_n$; and that there is a basic common term with incommensurable meanings distinguished as c_T, $c_{T'}$. Thus T is a function of $c_T; c_1, c_2, \ldots c_n$; and T' is a function of $c_{T'}; c_1, c_2, \ldots c_n$, or, to write this more formally,

$$T = T(c_T; c_1, c_2, \ldots c_n)$$
$$T' = T'(c_{T'}; c_1, c_2, \ldots c_n)$$

Evidently c_T must be combined meaningfully with $c_1, c_2, \ldots c_n$ (otherwise T would not be a meaningful theory). Likewise $c_{T'}$ must be combined meaningfully with $c_1, c_2, \ldots c_n$. Now if c_T, $c_{T'}$ can be combined meaningfully with a certain set of terms, then c_T, $c_{T'}$, must be commensurable. But this contradicts our supposition. Hence for c_T, $c_{T'}$ to be incommensurable, there can be no basic common terms with common meanings $c_1, c_2, \ldots c_n$. That is, all the basic terms of T, T' must be incommensurable. QED. The significance of this is that exponents of the incommensurability thesis almost certainly had the 'soft thesis' in mind and might not have seriously entertained it initially had they realised that it commits them to the 'hard thesis'. (I am not concerned here with ways in which this position may be evaded.)

Notes

1 Grünbaum (1966, 1969) has argued in detail that it is possible in principle to pin the falsity firmly on the theory, but it is not clear how we can know a priori that no surprise will turn up affecting the initial conditions and saving the theory. In any case he (Grünbaum, 1969, pp.1089, 1092) has granted that what he establishes is a 'well-nigh' conclusive result, that we can obtain falsification of a theory 'to all intents and purposes of the scientific enterprise' 'although we cannot falsify . . . beyond any and all possibility of subsequent rehabilitation.' This was never in doubt, and does not help to dispel the metascientific

force of conventionalism, which takes off from the unfalsifiability of a scientific theory *in principle*.

2   In other branches we get the same thing, for example, with the development of Hamilton's non-commutative algebra. A very simple illustration from elementary algebra is to be found in fractional indices; thus, $n^{1/2}$ has no meaning until it is given one by convention (which is done by making it obey the same rules as ordinary indices follow).

3   I get the impression that though Feyerabend in his recent work will not commit himself to realism, he has considerable reserve about conventionalism — one feels he regards it as irrational! In this matter I have tried to go further and take more risk than he.

4   This question was raised by my students Mr Kenneth Mantle and Mr Shalom Lappin at York University, Toronto, in 1969—70.

5   Professor J.L. Synge (personal communication) has hinted to me that no account of science can be satisfactory if it fails to do justice to the excitement of scientific enquiry. Presumably fitting jig-saw pieces together, reshaped by convention, to form a pattern would hardly help to explain the scientist's excitement.

6   I am indebted to Dr Mildred Bakan for pointing out that the effect on Tarski's theory is not central to my thesis.

7   Conventionalism would have to assume a form not itself undermined by the notion that observations are theory-laden, and it would have to come up with a conventionalist account of the truth-value of test observations (thus the reading of a voltmeter made to test Ohm's law would have to be accorded truth in a conventionalist sense); conventionalism would have to provide an account of observation-statements subject to the restriction that no term, not even an ostensibly observational one, could denote. Separately these requirements are insuperable, but conceivably a bold attempt might be made to cope with both together.

8   My colleague, Professor J.N. Hattiangadi, has already introduced a similar sort of distinction in connection with meaning.

9   Provided these might not have taken place, that is, subject to Popper's falsifiability conditions which are assumed here but not explicitly stated because not the focus of the present investigation.

10 Except in a world where there is no change — when we should not be alive and discussing the matter. Incidentally even Parmenides did not deny *all* change.

11 These tests are not of course theory-free but may be independent in the relevant respects of the two rival theories being tested.

References

Agassi, J. (1966), 'Sensationalism', *Mind*, 75, pp.1–24.
Ajdukiewicz, K. (1932), 'Das Weltbild und die Begriffsapparatur', *Erkenntnis*, 4.
Feyerabend, P. (1962), 'Explanation, Reduction and Empiricism', *Minnesota Studies in the Philosophy of Science*, Vol.3, ed. H. Feigl and G. Maxwell, University of Minnesota Press, Minneapolis.
Grünbaum, Adolf (1966), 'The Falsifiability of a Component of a Theoretical System', in *Mind, Matter and Method*, ed. P. Feyerabend and G. Maxwell, University of Minnesota Press, Minneapolis.
Grünbaum, Adolf (1969), 'Can we Ascertain the Falsity of a Scientific Hypothesis?', *Studium Generale*, 22, pp.1061–93.
Hattiangadi, J.N. (1974), 'The Importance of Auxiliary Hypotheses', *Ratio*, 16, pp.115–20.
Jarvie, I. (1964), *The Revolution in Anthropology*, Routledge & Kegan Paul, London.
Körner, S. (1964), 'Deductive Unification and Idealisation', *British Journal for the Philosophy of Science*, 14, pp.274–84.
Kuhn, T.S. (1962), *The Structure of Scientific Revolutions*, University of Chicago Press, Chicago.
Lakatos, I. (1968), 'Criticism and the Methodology of Scientific Research Programmes', *Proceedings of the Artistotelian Society, 68*.
Lakatos, I. (1970a), 'Changes in the Problem of Inductive Logic', *London Colloquium on the Philosophy of Science*, North-Holland, Amsterdam, Vol.3.
Lakatos, I. (1970b), 'History and its Rational Reconstructions', Philosophy of Science Association, Boston, October.
Nowell-Smith, P. (1971a), 'Cultural Relativism', *Philosophy of the Social Sciences, I*.
Nowell-Smith, P. (1971b), 'What Actually Happened', Lindlay Lecture, University of Kansas.
Poincaré, H. (1952), *Science and Hypothesis*, Dover, New York.
Poincaré, H. (1958), *The Value of Science*, Dover, New York, Part 3.
Popper, K.R. (1959), *The Logic of Scientific Discovery*, Basic Books, London and New York.
Quine, W.V. (1969), *Ontological Relativity and Other Essays*, Columbia University Press, New York, p.50.
Skolimowski, H.S. (1967), *Polish Analytical Philosophy*, Routledge & Kegan Paul, London, pp.145f.
Tarski, A. (1956), 'The Concept of Truth in Formalized Languages', *Logic, Semantics, Metamathematics*, Oxford University Press, New York.

Wisdom, J.O. (1970), 'Observations as the Building Blocks of Science in 20th Century Scientific Thought', Philosophy of Science Association, Boston, 23 October, *Boston Studies in the Philosophy of Science*, Vol.8, ed. Cohen and Wartoffsky, Reidel, Dordrecht, 1972.

Wisdom, J.O. (1971), 'Four Contemporary Interpretations of the Nature of Science', *Foundations of Physics, 1*.

Wisdom, J.O. (1974), 'The Incommensurability Thesis', *Philosophical Studies, 25*, pp.299–301.

Wittgenstein, L. (1921), *Tractatus Logico-Philosophicus*, Routledge, London.

15 Scientific theory: empirical content, embedded ontology, and weltanschauung

It is almost a commonplace that philosophers do not know what is the nature of metaphysics. But science? Here at least, it is generally considered, we are on safe ground; surely we all know what it consists of, at least in broad terms no matter what school of thought we follow. On the Bacon-Mill thesis it consists of observations and generalisations induced from them; according to the logical positivists' principle, it consists of observations and generalisations that are verifiable; or on Popper's demarcation criterion, science is what is empirically testable by observation. Thus, broadly speaking, science consists of observations and laws closely related to them. However, it is doubtful whether the contents of science are known, because a basic question arises about its components and boundaries. For it transpires that there are parts of science that are not related to observation, even indirectly. This conflicts with all three of the above approaches. It conflicts as strongly with Popper's as with the others; for it transpires that there are in his sense untestable intrinsic components. On his approach the problem then arises of how to assess the acceptability of the untestable parts of science, and for all the approaches whether the existence of ingredients of science that lack evidence or support leaves science without a rationale. My thesis is that science (as we have known it for 400 years) consists of three components, two of them hardly recognised — noticed indeed as influences, but not recognised as warp and weft; and I want to show that they have peculiar implications for the status of scientific

---

Published in *Philosophy and Phenomenological Research*, 1972, *33*, pp.62—77.

knowledge, as falling short of the support science normally demands, but that nevertheless they have a basic role in constructing solutions to problems.

Empirical content

Basic for my thesis is the notion of empirical content. Intuitively it is easily grasped and it is familiar to every scientist. Ostensibly the empirical content of a scientific theory is what the theory tells us, the information it provides, about the world. But, as we shall see, theories tell us certain things that are about the world and yet smack in some way of being non-empirical — that is to say, they go beyond what can be checked by observation. Here we find the notion emerging that empirical content is what is testable by sense-experience.

There has, of course, been no unanimity about this criterion or its interpretation. Inductive empiricists have regarded empirical content as built up out of observations; logical positivists as what is verifiable; Popper as what is falsifiable. My concern is with the problem of disproof in connection with the untestable components of science but my thesis is independent of whether or not one accepts Popper's criterion. The relationship between the two may be brought out in the following way.

Popper (1959) has provided a demarcation criterion to decide whether a general theory (understood in the broad sense of a theoretical system) belongs to science or not. It depends upon whether the theory is falsifiable or not. That is to say, a theory is scientific if it is possible to specify what observation[2] would refute the theory should it happen to be false. Although I hold that Popper has thus provided a superb criterion for empirical content,[3] this does not have to be assumed in what follows. My thesis requires, not that falsifiability is a *criterion* for empirical content (though I happen to think it is), but only that falsifiability is a *property* of empirical content. Alternatively, if one does not wish to characterise empirical content in this way, the present thesis concerns the difference between those parts of science that are falsifiable[4] and those that are not.

My thesis is that as well as empirical content, science contains other intrinsic components, which I hope to specify, but which his criterion was not designed for, and which are *not* falsifiable like empirical content.

To approach this, consider an example of empirical content, say that of the Newtonian law of gravitation, $F = M_1 M_2 / r^2$. From this (together with the other appropriate laws and especially initial conditions) observational consequences, such as the motion of the tides, can be deduced. If actual observations conflict with such consequences, the theory is refuted. And the motion of the planet Mercury does refute

Newton's law. Other examples of content of laws would be Galileo's law for the height of the falling bodies, $s=\frac{1}{2}gt^2$; or Kepler's law that the planets go around the sun in ellipses; which can be applied to the real world and could be refuted by observational means.

I would sum up the property in Popper's criterion as refutability by means of observation.[5] In other words, *the empirical content of a scientific theory is observation-refutable.*[6] This way of putting it will lead up to the next component of scientific theory.

Embedded ontology

Hard-headed scientists may think of science as consisting solely of its empirical content — all else being a blemish to be liquidated. My task is now to show that scientific theories have ontologies associated with them, in fact actually shot through them. The first part of my thesis is this. These ontologies are not refutable by observation: thus they carry the awful overtones of old-fashioned metaphysics about them. So it is at least important to examine scientific theories to see whether this is so or not, whether or not they do carry an embedded ontology. If so, the second part of the thesis concerns a method for assessing such ontologies. (There is a further thesis about an additional source of ontology, which will be explained later.)

I shall now consider a number of examples (see Wisdom, 1963).

Take the general law of conservation. If you say the sum of kinetic and potential energy is constant, that is refutable in Popper's sense. And it has in fact been refuted. You then make a new conservation law involving heat energy. This again is falsifiable — and has in fact been falsified. You continue to make new conservation laws with a new form of energy added to the old ones to preserve the constancy. That is quite in order; it is scientific. When a *specific* conservation law is found to be false, it is replaced by another *specific* one. But the *general* conservation law is in a different position. It is very hard to do without postulating that there is always energy in some form to make the total energy constant. But this is untestable; there is no observation, as was pointed out by Poincaré (1952), that could refute it in its most general form. For, if you find that there is evidence against a specific conservation law, yet you cannot find a new form of energy to add to preserve the constancy, the general law means that if we are lucky tomorrow we shall find some new form of energy that will do this; and you can never know that this will not happen. In other words, the general law is not observation-refutable.

Again a principle that has animated physiology (and rightly so) is that all bodily changes are due to physical causes (Wisdom, 1956, cf. Cohen, 1931, p.141). You will not find this written down in books, but you can suppose that it is written in invisible letters over all physiological laboratories. If a physiologist did not subscribe to it he would be out of his job or he would not be given a grant for his research. This principle may sound awfully thin and uninteresting, but if you think back a few hundred years to what it replaced, you will see that at one time it was of great significance. If you had warts, for example, you asked 'Who done it?' for you suspected an unfriendly witch. Nowadays, you ask 'What done it?', for you suspect a virus. This was an important change in the history of thought. But what I want to point out about even the new principle is, that it is not refutable by observation. If you find some bodily change for which you cannot track down a physical cause, you may take the line that a physical cause will turn up sometime, if not in the next 10 or 20 years then in the next 100 years, and if not in 100 years then later. Thus there is no way of specifying what would be allowed as a refutation. So this principle is not observation-refutable. Within the same family is that all mental changes are due to physiological causes (Wisdom, 1954). Perhaps a more interesting example would be that energy-values occur in all degrees. Again, this is not written down in any book on physics, but it certainly is presupposed by all classical physics, Newtonian and the various kinds of physics concerned with kinetics that preceded it. No one would ever have thought of writing it down as part of physics, for it is only with the advent of quantum physics that it becomes interesting; but it is a presupposition of classical physics nevertheless. Now if a certain energy-level *is not* to be found, there is no way of showing that it *could not* be found — it may turn up tomorrow. Hence this part of classical mechanics is not refutable by observation.

I would call this an example of a continuity-principle.

Another instance of the continuity-principle is that matter is infinitely divisible — again not observation-refutable.

Another example is that space is infinite. There is no way of refuting this by observation. Yet it is a presupposition of Newtonian mechanics.

A highly interesting example is that space is absolute — again not observation-refutable. The cash value of Newtonian mechanics for hard-headed scientists is the empirical content of $F=M_1 M_2/r^2$. But Newtonian theory essentially involves both empirical content and the absoluteness of space. That space is absolute is not written down in the deductions, but it permeates them. For you cannot carry out any of the work of Newtonian mechanics without supposing bodies are absolutely rigid; as you move a measuring rod from here to there it stays rigid, that is, stays the same length in a classical sense. This may be taken to be one form of the absoluteness of space. Another interpretation would be that velocities are compounded according to the ordinary algebra of vector addition.[7] So the introduction of the irrefutable idea of the absoluteness of space is not just an aberration of

Newton's but a presupposition that is actually, even though covertly, used in Newtonian mechanics.

Some of these examples are confirmable but not refutable; others neither.

Thus empirical content is not pure: it contains what may be called an *embedded ontology*.

Premisses, inference, and refutation

A theory as a whole thus involves a premiss about empirical content plus another (ontological) premiss which perhaps is better not stated as a separate premiss, because usually in logic premisses are taken to be independent of one another. However, the one about absolute space is impregnated in the one about content. And this fact has a logical implication of significance.

Suppose now you get an observation-refutation of, say, Newtonian mechanics, and let us agree that the premisses consist of the empirical content of $F=M_1 M_2/r^2$ and the absoluteness of space. An observation-refutation means that some conclusion is false, and therefore in a valid inference that one of the premisses is false (possibly both but certainly one). But you do not know *which* premiss is refuted. Could we get a refutation of the premiss about absolute space with the content left untouched? This would seem to be impossible. Clearly if the premisses are independent of one another (in the sense that neither follows from the other) and a false conclusion is obtained (the inference being valid), any one of the premisses may in general be false, and there may sometimes be a reason for judging one of them to be false rather than another. Why not here? Given that the ontological premiss pervades the content, it is possible for the content to be false and the ontology true. When Newton's theory was refuted by the observation of Mercury, it would have been possible to replace it with a theory having a different content yet retain the ontology of absolute space. But the converse is not possible: you cannot give up that ontology without altering the content and therefore giving it up too; that is, you cannot give up the ontology and retain the content. That is to say, the ontological premiss cannot be false and the content true. Hence an observation-refutation of Newtonian mechanics can only refute the content but not the ontology.[8] This result is not, of course, news: metaphysicians have held for centuries that ontology is not refutable by observation. My aim has been to show that ontology is part of empirical science and it is not refutable in Popper's sense.

We are therefore faced with a highly interesting problem: how is the validity of the embedded ontology of science, which is not testable by observation, to be decided?

Before entering upon this, a brief excursion into cognate matters is desirable.

Theory-refutability

Consider an empirical theory that is (a) observation-testable, (b) observation-tested, and (c) corroborated (where 'corroboration' is short for 'failed falsification by the best available tests'). In short, it is testable, tested, and corroborated. Until it is overthrown, that is, observation-refuted, its logical consequences have to be accepted and anything that is incompatible with its logical consequences has to be rejected. If relativity is assumed to be such a theory, it is a logical consequence that space is relative, which is incompatible with the Newtonian ontology of absolute space. This ontology is therefore refuted by a scientific theory which is testable, tested, and corroborated. Thus the Newtonian ontology, though not observation-refutable, is *theory-refutable*, and in fact theory-refuted. Again, suppose that quantum mechanics is testable, tested, and corroborated. It is a consequence of quantum mechanics that there are eigen-values of energy, that is, that in certain contexts energy exists only at discrete levels. In other words, energy does not occur continuously at all levels. This consequence of quantum mechanics is inconsistent with the presupposition of all classical physics that energy can exist at all levels. Thus this piece of ontology embedded in classical physics, though not observation-refutable, is theory-refutable, and in fact theory-refuted (Wisdom, 1963).[9]

This conclusion goes much further than merely making the point that theories can be inconsistent with one another: what is asserted is that a metaphysical type of theory, that is, the embedded ontology, notoriously irrefutable and certainly not refutable by observation, is found to be refutable by an empirical theory.[10]

The great advantage of this procedure of using theories to refute is that it gives a way of handling various features of science for which metascientists have been unable to find a satisfactory method of criticism.

It has a curious application.

If we go back to preclassical physics, we find that in astronomical theory the earth was at the centre of the universe, and this was explained on the basis of some value-judgement connected with man, such as that his value status necessitates his being at the centre. This was refuted by the heliocentric theory. Thus even a value-judgement that involves the physical world can be theory-refutable.

Non-absoluteness of refutation

It will be noticed that an embedded ontology that is theory-refuted is only *conditionally refuted*. Once an ontology is refuted it is not refuted for all time, because clearly the refutation is dependent on the corroboration of a given theory, and that theory may not last, for it may, however well attested, like Newton's, turn out to be false. Then the refutation it provides collapses. And the refuted ontology becomes once more a candidate for acceptance. This puts the method of refutation in a somewhat weaker position apparently than does Popper's method of observation-refutation for empirical content. And this is so, up to a point — but only up to a point. It is well known that even an observation-refutation is not absolute; Popper explicitly states that even a false consequence is not an *absolute* refutation of a theory. The intuitive broad idea is that a false consequence refutes a theory, and that one false consequence refutes a theory absolutely and for always. But the false consequence is in its turn hypothetical, open to challenge, and might turn out to be in error. Then the theory, taken to have been refuted, would once more become a candidate for acceptance. Hence rejection by theory-refutation is hardly overwhelmingly shakier than rejection by observation-refutation. But it is considerably shakier, for an observation is *less* corrigible than a (high-level) theory (noting, as Popper says, that an observation-statement is itself a low-level theory).[11] The difference may be seen in the fact that theory-refutation involves refutation at two levels, one being refutation by the theory, the other being possible refutation of that theory by observation.

Ontology non-corroborable

We have that the empirical content of a scientific theory is observation-refutable, and that the embedded ontology is theory-refutable. In addition empirical content is corroborable, or observation-corroborable, and when possible falsification by observation fails it is corroborated. To obtain corroboration of empirical content by specifying a possible observation-refutation involves (i) specifying a possible observation forbidden by the theory within definite limits of accuracy, (ii) the high probability of its occurrence relative to other theories, and (iii) its actual non-occurrence. By contrast, embedded ontology is not theory-corroborable; for, when a new observation-corroborated theory is inconsistent with the ontology, that ontology is refuted, but failure of the new theory does not corroborate it. More explicitly, to obtain corroboration of embedded ontology by specifying a possible theory-refutation would involve (i) inventing a new theory (in sufficient detail for testing to be possible) before it is discovered, or rather at will or when need arises, (ii) the high probability of its being corroborated, and (iii) its actually being refuted. Thus while with empirical content we can specify what new observation, if it occurred, would refute it, *we cannot specify what new theory, if it could be found, would refute the ontology* — embedded ontology is non-corroborable.

Thus scientific theory is a compound of corroborable and non-corroborable ingredients. It is not sufficient to treat the compound as a single whole, a refutable theoretical system, because an observation-refutation can test (refute or corroborate) the empirical content but must leave the embedded ontology untouched. It is a question, therefore, whether there is a rationale for acceptance of the embedded ontology of science and therefore even of science itself.

Weltanschauung

The last part of my thesis is that some ontology, so far from originating along with the empirical content, usually arises independently of it and comes from a radically difference source.

An ontology makes an assertion about the entities that populate the universe of discourse or about relations between them.

It is possible to interpret an ontology as an expression of policy or programme. This is overt with 'for all physiological changes seek only physical causes'. Such a policy tells you what kind of entity will count as part of a scientific answer and what will be disallowed. Even in its programmatic form, therefore, it implies an ontology. But it is not an embedded ontology. For it would be possible to deny this ontology and give it up as a general programme without, for example, giving up the innumerable discoveries that have been made in physiology. In fact, many such discoveries have been made on the basis of this programme, even though it may have prevented others from being made. Such an ontology, however, is, like an embedded ontology, non-observation-refutable.

This leads on to the general consideration of policy/method, or, if you prefer, metascientific postulates. What counts as empirical science is not uniquely characterised by its greatest practitioners. Not only did Galileo, Descartes, and Newton have different conceptions of science from one another, but they had different conceptions of it from what *we* do in the twentieth century. These conceptions tell you what is to count as being scientific (as regards theories and as regards data), what the strategy is of doing science, or what the world consists of. Thus underlying the conception of method lay a policy of the scientist that formed part of his science by setting various limits to what science is.

The mathematical or the experimental conceptions involve the policy that certain studies cannot be scientific. The conventionalist, pragmatic, operationalist, instrumentalist, or observationalist conceptions exclude from science any claim to describe the world, and exclude any dealings with the furniture of the earth (thereby disallowing certain questions from being asked within the confines of science) (cf. Cohen, 1931, p.142).

Thus the policy/method can prescribe ontology — in these cases negatively, that is, proscribing ontology. Science, by its policy/method, can also prescribe the contents of its ontology (cf. Cohen, 1931, p.142; Watkins, 1958, pp.355—6, 358).

The general conception of science at any one time or in a dominant research group constitutes a 'set' of the scientist which prescribes certain features of science. It imposes a framework and legislates type of content. No English expression such as 'point-of-view' or 'general outlook' is strong enough, 'world-view' has overtones more mystical than required, and 'scientific ideology' could be sadly misleading, so I propose to use the German word 'weltanschauung'. But I shall also use the terms 'framework' and 'schema', simply as synonyms where they would fit the context more smoothly.

The question now arises whether the ontology prescribed by such a weltanschauung (for example, atoms, viruses, and so on) can be subject to any kind of testing. It is already non-observation-refutable. Is it theory-refutable?

For the most part, such matters are discussed as issues in metascience (methodology, philosophy of science), and the trend of the argument amounts on the one hand to the contention that ontology cannot be settled 'scientifically', for the most we can vouch for 'scientifically' are our observations, and on the other hand to the contention that certain significant questions are liquidated (for example, why a radium atom breaks down). Nonetheless it is not beyond the bounds of possibility to obtain theory-refutation, for if we could obtain a theory, testable, tested, and corroborated, about the ontology proscribed by the functionalist conception, we should have to reject this weltanschauung. An example would be: the policy-method that explanation must be given in terms of action-by-contact would exclude Newtonian celestial mechanics from science. Thus a weltanschauung would be rejected by an empirical theory that successfully refuted the ontology prescribed by the weltanschauung.

Let us now assemble the various components of a scientific weltanschauung.

Along with the embedded ontology of a theory, such as absolute space in Newtonian mechanics, there develops a weltanschauung. That is to say, one sees the world in terms of absolute space or whatever. The bare empirical content, I think, does not contribute to this.

Thus we have the two components: the attitude arising from/with embedded ontology and the policy underlying the conception of the method of science; but these in their turn generate a third — the emotional charge of a weltanschauung. I would add, further, that the weltanschauung of policy/method prescribes the *essence* of science for a period. A weltanschauung is an integral part of science because it determines what shall and what shall not form part of its ontology. In line with this notion of policy prescription, Lakatos (1968) has put forward an interpretation of research programmes which constitute an important new element in the way a weltanschauung works.

It may be noted in addition that, though a weltanschauung generates an ontology, it may on occasion lead to new empirical theory and also the converse, that a new ontology and a new empirical theory may generate a new weltanschauung. This chapter, however, is hardly concerned with the sources of weltanschauungen.

Forms of metaphysics, the paradigm, and decidability

The present thesis has certain corollaries.

It is not enough to refer to the 'metaphysical' ingredient of science, as has usually been done. Several different things can come under this heading (I would say there are three radically different kinds of metaphysics). One of these and only one is equivalent to the embedded ontology of science.

It seems to me that Kuhn's (1962) conception of a paradigm is in the main to be associated by him with the *content* of a dominant theory (there are indications to the contrary, but since he did not anticipate the categories of the present chapter he was naturally not explicit on the point). In my view, the paradigm is a genuine and important concept, but should be associated with the embedded ontology and with the policy/method of the weltanschauung.

I wish to suggest that high scientific controversy concerns not content but embedded ontology and weltanschauung (the ideology of science at some period). If empirical content dominates, we may expect to settle the matter before long with an empirical test.

Sometimes, as in contemporary astronomy, the evidence is hard to get and we just have to wait patiently; that this has not been the attitude adopted is due, I suggest, to the fact that the two rival theories involve violently opposed weltanschauungen. But if faced with an ontology that cannot be refuted by empirical investigation, by observational means, by experiment, all that can be done is argue about it ineffectually, until a theory that can refute it happens to turn up. But until that happens we are left with an undecidable presumption. We are also in a position to understand Kuhn's contention that scientists imbued with different paradigms cannot communicate with each other. While I do not think this is in fact so, there is some truth in it — they *can* have difficulty in communicating. When confronted with a new empirical theory, such as relativity, one naturally looks at it in terms of the old weltanschauung.[12]

Thus empirical science consists of empirical content, the embedded ontology of the theory (embedded in the content), and the non-embedded ontology prescribed by the policy/method weltanschauung.

The possibility exists of criticising and even refuting the last two when they contradict empirical consequences of corroborated theories. Unlike empirical content, embedded ontology and weltanschauungen cannot be corroborated. While they are harmful if they are used to proscribe empirical content that might prove serviceable, and also if they are used to dogmatise about the furniture of the earth, they are a harmless adjunct to corroborable empirical content; hence their use is legitimate so long as the empirical content embodying them is corroborable.

Recent disquisitions on science and metaphysics

Many philosophers have written on the relation between science and metaphysics, but the present thesis goes well beyond the relationships that have become widely noted. Perhaps the most striking example from recent writings is to be found in Agassi's (1956) PhD thesis, where he described metaphysics as a 'programme' in discussing its influence on research:

> When trying to form a hypothesis, we are usually aided by a general view, some principles which are accepted prior to the framing of a hypothesis, and which prescribe some characteristics of the hypothesis we intend to form.
>
> The principles are accepted *a priori* in the sense that they are adopted prior to the hypothesis, which we intend to frame, and do not stand or fall with it: if one hypothesis which agrees with the principle fails, we can always try to frame another hypothesis which has the same general characteristics prescribed by the principles, but which may otherwise be very different from its predecessor.

Notable writings in connection with the present theme are Burtt (1924), Cohen (1931), Agassi (1956, 1962, 1964), Watkins (1958), Bunge (1959), Feyerabend (1962),[13] Lakatos (1968), and Körner (1970).[14] Burtt (1924) regarded metaphysics as an integral part of science; I have found that he even speaks of it as 'imbedded' (but I think he means the way in which an outlook pervades one's work, hardly in the sense of 'embedded ontology'). The notion of metaphysics as regulative is fairly common. According to the distinguished scientist, J.J. Thompson (1907), '. . . a theory of matter is a policy rather than a creed'. Popper (1959) held that it could be interpreted in terms of methodological rules. Hutten (1950) has mentioned it under the heading of 'policy-directive', and I (Wisdom, 1954) under the heading of 'counsel for research'. Like Agassi, Watkins (1958) described it as 'regulative'. This would also seem to be the view held by Bunge.[15] In addition, a distinctiveness of weltanschauungen has been noticed by Bromberger (1963) and by Jones (1970).

Since the distinction between embedded ontology and weltanschauung had not been drawn, it is often not clear under which heading writers would have placed metaphysics; but the above citations seem to indicate the latter.

Some of these philosophers came near to the idea of refuting metaphysics (here perhaps in the sense of ontology rather than weltanschauung) by means of a scientific theory, without quite making the point (Bartley (1962) came closest). Watkins (1958), using several of my examples, gave a nice treatment of the influence of metaphysics on science. Thus he showed that metaphysics can rule out certain scientific theories; which seems to me correct and important. He did not, however, consider the converse, that science can rule out metaphysics, which was the main point of my paper, 'The Refutability of "Irrefutable" Laws' (Wisdom, 1963), and elaborated here, in the form that a corroborated scientific theory can refute an ontology; on the contrary, Watkins referred to metaphysics as neither testable nor demonstrable nor refutable (Watkins, 1958, pp.363—5) — his aim, legitimate but different, prevented him from picking up a result almost at his feet.

As regards Watkins' contention, which with nicely handled examples he made more explicit than most authors, that (in my terms) a weltanschauung can prescribe and proscribe the type of science that shall be pursued, I am wholly in agreement; but I have given this a stronger form, namely that this phenomenon is not a chance side-effect of scientists' being all too human, but an intrinsic connection. Lakatos (1968) has added an important new element to the heterogeneous types of contents of a weltanschauung: he introduces the notion of a research programme as an extrapolation of modifications to shore up a theory against successive counter-examples — which evidently prescribes the type of theory that shall be adopted (and proscribes others). This is a powerful and hitherto unnoticed way in which a weltanschauung exerts its influence. Agassi (though he gives an actual example of theory-refutation and adds that the history of science is the history of new science elimating old frameworks) regards regulative ideas as being outmoded rather than refuted. Further, he gives striking examples of regulative ideas opening the way to new empirical theory. But basically, despite a faint tendency the other way, he regards those ideas as falling outside science. On my thesis they fall inside.

Agassi (as well as Popper) holds that ontology is *criticisable*, though in some sense short of refutable. One of my problems, however, is to understand the notion of criticism where known methods of refutation are inapplicable.

But the large-scale disparity between the present thesis and existing writings in the area renders detailed discussion of the literature scarcely relevant.

Conclusion

We are now in a position to sum up what characterises the present thesis. (i) The specific distinction between ontology and weltanschauung is here obviously central. (ii) The refutability of an embedded ontology by means of an empirical scientific theory had not fully emerged in the one or perhaps two writers who were concerned with a problem near to this one. (iii) While it was recognised that science had ontological features, it was not realised that, so far from being dispensable excrescences born of a primitive scientific outlook, they are ineradicable components embedded in all scientific theories. For it was overlooked that the embedded ontology is independent of the empirical content, whereas the empirical content is pervaded by that ontology. (iv) A further consequence is that the meaning of metaphysics, when a part of science, becomes specific. (v) A weltanschauung consists of the framework that arises out of an embedded ontology, the policy/method prescribed/proscribed by such a framework, and the emotional charge invested. A weltanschauung does not just vaguely suggest the future direction of science but prescribes its ontological make-up. (vi) And while it was widely recognised that weltanschauungen *influenced* scientists and their theories, this being conceived of as a human failing detachable from science proper, the present thesis holds that a weltanschauung is an unavoidable component intrinsically connected with empirical theory, because it determines what shall and what shall not form part of the ontological make-up of science. (vii) The notion of a paradigm becomes sharper. (viii) A weltanschauung may become rejected by an empirical theory that successfully refutes the ontology prescribed by the weltanschauung. (ix) The notion of criticising, where observation-refutation was deemed impossible, is now understandable; for an appropriate mode of refutation has been found. (x) The present thesis has markedly different implications for the understanding of the nature of science, knowledge, and rationality. (xi) The overall conclusion is that, whereas all metaphysics used to be regarded as a foreign body, lying outside science, some metaphysics is here treated as part of the nature of science itself.

Postscript

Although the literature of philosophy of science and history of science has long contained allusions to metaphysics, and of recent times has given more attention to it, there has been a lack of specificity. The allusiveness concerns both lack of specificity about the nature of the animal, metaphysics, and also about what metaphysics does and does not do.

The distinction between embedded ontology and weltanschauung enables us to clear up certain points. Thus ontology is unavoidably present in all science, even the most developed and respectable, though hardly admitted a place by the hard-nosed scientists. It is not amenable to 'ordinary' scientific testing, can never be corroborated, though with great luck it may on occasion be refuted. This does not constitute a bad blemish in science, nor render ontology a drag on discovery, for an ontology is permissive in a certain sense, in that, until refuted by a later theoretical development, it is part of a working exploration — but its acceptability is restricted to what the empirical content in which it is embedded can do in its name.

In natural science, writers wobble between identifying metaphysics with embedded ontology and with weltanschauungen, usually with overtones indicating the latter. Perhaps this affects the natural scientist unfavourably, because weltanschauungen play an even less obvious part in science. They have played an enormous part in the past, but this is forgotten or not realised. But they come in all the same, as for instance over the way to approach quantum mechanics. That exemplifies their bearing on natural science. In philosophy of science the notion bears on objectivity, incommensurability of theories, and cognate themes. But it is in the social sciences that its influence is most overt, and social scientists explicitly refer to it, under a variety of different names. One of the most important ideas that emerges is that, with an unprovable weltanschauung, science is the reflection merely of a point of view, and even that the weltanschauung distorts the results achieved. Such a conclusion results from not studying what weltanschauungen do and do not do. It is very clear that a weltanschauung does indeed reflect only a point of view. But what the weltanschauung effects is to promote what comes within the point of view and discourage what falls outside it: what is discovered in its name can be valid if satisfactorily tested; what is not investigated is a loss. But what is gained is not thereby invalid or distorted.

Thus the notion of a weltanschauung is basic for dealing with problems of certainty, objectivity, reality, truth, distortion, rationality.

Appendix 1: The Demarcation Problem

Disregarding the implied solutions to the demarcation problem to be found in the work of Bacon or of Mill and restricting ourselves to Popper, the solution he gave to this problem may be expressed in brief form by the following criterion: a statement is scientific if it has consequences that can be tested by observation. This criterion is, as I have indicated, equivalent to taking empirical content to be observation-refutable.

Since, as I have now made clear, there are other components of science which are not observation-refutable, the relevant one here being embedded ontology, it is clear that the line of demarcation drawn by Popper will not quite serve. Indeed, paradoxically enough, the criterion he gave hives off metaphysics into the realm of non-science, and yet some non-science in the form of metaphysics in the sense of ontology is, as I have shown, part of science itself. What is paradoxical is that the main ingredient that Popper wished to exclude from science has now been found to be one of its significant components.

It may indeed be that Popper would not disagree with the spirit of this insofar as he later pointed out the way in which science evolved historically out of metaphysics. This contention of his, which is no doubt true, does not necessarily conflict with his demarcation criterion but it would at least suggest a consideration of the need to draw the line elsewhere.

It does not seem to me that the required dividing line is very difficult to find. What I have shown is that embedded ontology, though not observation-refutable, is theory-refutable. This is the main form of metaphysics lying within the field of ontology that I propose to include within the corpus of science itself. Let me here make the broad new contrast. The great cosmological forms of ontology which have occupied philosophers for over 2,000 years deal with very abstract questions, for example, to do with substance, relationships between substances, the nature of universals, and so on. These forms of ontology, which constitute one main type of metaphysics, form no part of science. They may or may not be in their turn theory-refutable, but that is not the relevant feature here. What I am including within the corpus of science is a compound consisting of empirical content, to which Popper's criterion of observation-refutability applies, together with embedded ontology, to which my criterion of theory-refutability applies, provided it is noted that the ontology in question really does belong to the empirical content. The dividing line is thus drawn in such a way that embedded ontology falls within science and the great cosmological ontologies, which make no claim at all to have any relation to the empirical content of any scientific theory, fall outside the realm of science.

This is the simplified picture; it can be made slightly more realistic, though slightly more complicated, by including a reference to another form of ontology which is not embedded but which nonetheless does direct science in a highly significant way. It is not embedded in the sense of being a necessary part of the empirical content of a science but it happens to be part of some great corpus of scientific theorising. To remind the reader of the difference, absolute space in Newtonian mechanics is embedded ontology in the sense that the empirical content theories cannot be understood or used without the embedded ontology. On the other hand the principle that all bodily changes are due only to physical causes is a piece of ontology that is not embedded in biological or physiological theories, even though all the great classical developments in physiology conform to it. It was not part of their structure in the sense of part of the meaning. Now such ontologies are indeed part of the body of science although in a different sense. What they are amounts to being policies for research dictated by a weltanschauung which prescribes or proscribes what shall or shall not count as belonging to science.

The weltanschauung of science, however, belongs to science in a different way from the embedded ontology. It is a general framework imposed upon science which gives science its shape. It is not one of the recognised forms of classical metaphysics though it does prescribe a form of metaphysics or equally it may proscribe a form of metaphysics.

The upshot of this is that science is now seen to contain three parts, (i) the empirical content of a scientific theory, (ii) the embedded ontology, and (iii) the prescriptions and proscriptions of a weltanschauung. It may seem strange and lopsided to include within the body of science the weltanschauung of the subject itself as one of its own components. But I do not think that this attribution of a tiered structure is an objection, for each level or tier is subject to its own criterion. In this last case the weltanschauung is, like embedded ontology, subject to the criterion of theory-refutability. On this line of demarcation, then, what lies outside science would be ontologies or weltanschauungen which have no relationship to science itself.

This development of the demarcation criterion would seem to be in line with the spirit of Popper's own procedures and philosophy. In a narrow sense it contradicts his result insofar as this treatment asserts that his criterion, which holds for empirical content, is wrong because insufficient. But this is very different from asserting that his criterion is wrong because it is quite mistaken throughout. His criterion simply needs to be supplemented, though in a significantly different manner. Although this treatment is in a narrow sense a refutation it seems to me more properly to be regarded as an evolutionary development from his position, from which indeed it develops in the first place. And this especially because it arose in order to try to meet a difficulty or a deficiency which I had found in his excellent treatment of empirical content.

Appendix 2: The Notion of a Model

In recent years there has been an attempt to introduce the notion of a *model* more widely into philosophy of natural science. It also became customary in some economic circles to use it in theoretical economics.

There is no need to go into the variety of purposes that may have been involved in these discussions. Suffice it to remark that some observers have got the impression, in both fields, that a model was a theory, and some economists have suspected that it was used for no more reason than to avoid the concept of 'theory', which in their circles seemed to smack of barren speculation.

It is possible that the difficulty that was experienced in locating the reference of, or a suitable reference for, the notion arose because a distinction that might have smoothed the path had not been explicitly drawn. I would suggest that it would be viable to direct the reference not to the *content* of a theory but to its *ontology*. Thus we could reasonably take the notion of an ideal gas described in terms of perfectly elastic billiard balls to be an ontological model.

However, we might also take the notion of wave or particle to be a weltanschauung prescribing the approach to be made to fundamental theory. It would depend on context whether a model constitutes an ontology or a weltanschauung; but for the most part perhaps the notion of a model would be ontological.

Notes

1 A sketch of this chapter, about a third of its length, was written for the International Congress at Vienna (Wisdom, 1969). The main ideas were applied in two different contexts in 1966 (Wisdom, 1971, 1974), but were not abstracted from those contexts. This was done sketchily in the above paper and much more fully here.
2 Strictly, Popper formulates in terms not of observation but of an intermediary, which is itself falsifiable by observation (this also can be more strictly presented), so that by transitivity it comes to the brief version given above; the stricter amplifications are not relevant in the present context.
3 Popper (1959, sec.21, p.86) identifies 'empirical' with 'falsifiable'.
4 Recently with a revival of conventionalism by Kuhn (1962) and by Lakatos (1968, 1970*a*, 1970*b*), the falsifiability thesis has been repudiated on the grounds that a scientific theory or its empirical content cannot be falsified and that it is not part of the structure of research to falsify. This is, of course, irrelevant to the present thesis, which is concerned with the relative falsifiability of empirical content (whether by deliberate aim or by metascientific reconstruction) and the unfalsifiability of other content by the same empirical method.
5 This is not a rigorous formulation but it suffices to bring out the

most central part of his criterion, though so far as the present thesis is concerned it leaves a trivial hole unplugged.

6   Agassi (1964, pp.198, 200) also has pointed out that science contains more than empirical content, and that it is the empirical character of science to which Popper's criterion applies. (He attributes this to Popper in unpublished lecture courses.)

7   From these interpretations it is to be understood that the contrast is with relativistic (Einsteinian) space and not with relative (Leibnizian) space. So far as *these* properties are concerned, of rigidity and of vector addition, Leibnizian space is not relativistic. We do not have to go into the question whether Newtonian space presupposes an absolute centre, and Leibnizian relative space dispenses with one. This example is, of course, controversial, and to this extent unsuitable to my purpose, but it is so convenient an example that it would be a pity to exclude it.

8   At least this is a strong presumption. I cannot prove it, however, so the possibility should be kept open that an alternative ontology could be devised to fit the content, in such a way as to lead to the appropriate predictions including the previously discrepant ones.

My former student, Dr Catherine May (1977), has given some able criticism and critique of my proposals about embedded ontology and weltanschauung. Dr May maintains, contrary to the above chapter, that a later false theory whose ontology is incompatible with the ontology of the earlier theory, in thus failing to falsify the old ontology provides corroboration of it. I am unable to see this. The failure of the new theory would (I think) leave the old one precisely where it was, with some explanatory success, some dubious observational features, and *ontology untouched*. Dr May's position is rich enough, however, to allow the possibility of being worked up into a viable thesis.

9   Bartley (1962) has nearly made the same point: he utilised similar ingredients, but was aiming at a different target. Watkins (1958, p.358) arrived at a correct conclusion with a family resemblance to mine, but he had a different aim and made in fact the *converse* point.

10   My former student, Mr Thomas K. Bowers (in 1972), has made an interesting use of the distinction between empirical content and ontology against, for instance, Lakatos' (universalised) theory of research properties: the succession 'of revised but distinct theories within a particular scientific tradition, without challenging the ontology which constitutes that tradition, is precisely attributable to the falsifiability of some consequences of scientific theories'. (Unpublished)

11   Popper's broad and clear theory of refutability is one thing. Its sophisticated elaborations and/or defences against subtle queries are another. Broadly, an observation is always questionable; otherwise expressed an observation is fraught with interpretation; otherwise expressed an observation-statement is *theoretical*. So one *could* say

that, for Popper, empirical theories are in fact, not observation-refutable, but theory-refutable. No Popperian would, of course, urge this as an *objection* to my thesis (though he would be alive to the point), because Popperian doctrine is in any case to attribute to a view not its overt expression but its strongest possible form. This could be done — at the expense of complicating this chapter — by writing of observation-statements that they describe not observations but theories, so that my thesis would be transformed into the following: that Popper's theory of empirical content is refutability by observation-level theories, while I am adding a theory about embedded ontology which is not refutable by observation-level theories but by high/higher-level theories — this new level is an ingredient of science in any case, whichever way it is expressed. Although I personally think Popper is right that observation-statements are theoretical (in a modest sense), I do not think these are the most appropriate terms in which to conduct the exegesis of my distinction.

12 I recall that when I was an undergraduate there were still those about who felt the revolution in modern science constituted a *treachery* to Euclid.

13 Here and there Feyerabend shows himself alive to the influence of *general outlook*. But he makes no attempt to follow up what I have called weltanschauung. Here, too, as with other points of common interest, I have tried to pursue the matter some little distance, and have suggested grounds of 'rational' evaluation of a weltanschauung.

14 Körner (1948, 1969, 1970) has put it forward as a programmatic view of metaphysics.

15 A marvellous example of the interplay between quantum mechanics and the philosophy of quantum mechanics constitutes the core of a work done by Landé (1965). Another is Sabra's (1967) work on philosophical problems in the history of science. Professor A.I. Sabra's contributions in seminars some thirty years ago, which later became the work referred to, first aroused my interest in the present area.

References

Agassi, Joseph (1956), 'The Function of Interpretation in Physics', PhD thesis, University of London, pp.9f.

Agassi, Joseph (1962), 'The Confusion between Physics and Metaphysics in the Standard Histories of Science', *Ithaca*, p.233.

Agassi, Joseph (1964), 'The Nature of Scientific Problems and Their Roots in Metaphysics', *The Critical Approach to Science and Philosophy*, ed. Mario Bunge, New York.

Bartley III, W.W. (1962), *Retreat to Commitment*, New York, pp.85, 159.

Bromberger, Sylvain (1963), 'A Theory about the Theory of Theory and about the Theory of Theories', *Philosophy of Science, The Delaware Seminar*, Vol.2, ed. Baumrim, New York and London.

Bunge, Mario (1959), *Metascientific Queries*, Springfield, Ill.

Burtt, E.A. (1924), *The Metaphysical Foundations of Modern Science*, London, pp.20, 224.

Cohen, M.R. (1931), *Reason and Nature*, New York, 1964 edition, p.138.

Feyerabend, P.K. (1962), 'Explanation, Reduction and Empiricism', *Minnesota Studies in the Philosophy of Science*, Vol.3, ed. Feigl and Maxwell, University of Minnesota Press, Minneapolis, pp.28–97.

Hutten, E.H. (1950), 'Induction as a Semantic Problem', *Analysis, 10*, pp.126–36.

Jones, W.T. (1970), 'Philosophical Disagreements and World Views', *Proc. Am. Philos. Assoc., 43*, pp.24–62.

Körner, Stephan (1948), 'The meaning of some metaphysical propositions', *Mind, 57*, pp.275–93.

Körner, Stephan (1969), 'Categorial change and philosophical argument', Proc. Israel Academy of Sciences and Humanities, *3*; and (1970), *Categorial Frameworks*, Oxford.

Kuhn, T.S. (1962), *The Structure of Scientific Revolutions*, University of Chicago Press, Chicago.

Lakatos, Imre (1968), 'Criticism and the Methodology of Scientific Research Programmes', *Proc. Arist. Soc., 68*.

Lakatos, Imre (1970a), 'Changes in the Problem of Inductive Logic', *London Colloquium on the Philosophy of Science*, Vol.3, North-Holland, Amsterdam.

Lakatos, Imre (1970b), 'History and its Rational Reconstruction', Philosophy of Science Association, Boston, October.

Landé, Alfred (1965), *New Foundations of Quantum Mechanics*, Cambridge University Press, Cambridge.

Poincaré, Henri (1952), *Science and Hypothesis*, Dover, New York, Ch.8.

Popper, K.R. (1959), *The Logic of Scientific Discovery*, Hutchinson, London and New York.

Sabra, A.I. (1967), *Theories of Light from Descartes to Newton*, Oldbourne, London.

Thompson, J.J. (1907), *The corpuscular theory of matter*, Constable, London, p.1.

Watkins, J.W.N. (1958), 'Confirmable and Influential Metaphysics', *Mind, 67*, p.345.

Wisdom, J.O. (1954), 'Is Epiphenomenalism Refutable?', *Proc. of the 22nd Inter. Congress for the Philos. of Sc.*, Zürich, published 1955.

Wisdom, J.O. (1956), 'Psycho-Analytic Technology', *Brit. J. Philos. Sc.* 7, reprinted in *Psychoanalytic Clinical Interpretation*, ed. Paul, New York, 1963, p.158.

Wisdom, J.O. (1963), 'The Refutability of "Irrefutable" Laws', *British Journal for the Philosophy of Science, 13*, pp.303—6.

Wisdom, J.O. (1969), 'Scientific Theory: Empirical Content, Ontology, and Weltanschauung', *Proc. XIVth Inter. Congr. Philosophy*, Vol.III, Vienna, pp.147—53.

Wisdom, J.O. (1971), 'Freud and Melanie Klein: Psychology, Ontology, and Weltanschauung', *Psychoanalysis and Philosophy*, ed. Hanly and Lazerowitz, International Universities Press, New York, pp.327—62.

Wisdom, J.O. (1974), 'The Nature of "Normal" Science', *The Philosophy of Karl Popper*, Library of Living Philosophers, ed. Schilpp, Open Court, Chicago, written in 1966, reproduced in Chapter 4 above.

16 Weltanschauungen as sources of our knowledge and our ignorance

Where is the trouble spot of fundamental problems? If you are a Baconian you will say you lack data; if you are a Popperian you will say you lack ideas or theories. At this stage I take it that the first is a dead loss. While following the second, I add the qualification that it is not enough. Why are problems so intractable or why is it so difficult to develop even passable theories? Is it basically that the intrinsic difficulty is so great?

Having noticed that weltanschauungen influence our theories, the question began to raise itself whether there might not be some specific influence — notably weltanschauungen — blocking our efforts to find and work on new theories.

Although this is a psychological phenomenon, it is with the epistemological position that I am concerned: there can be a logical incompatibility between a weltanschauung and a new theory. Here the aim is to explore the epistemological relationship between weltanschauungen and theories with empirical content, with special focus on the phenomena of discovery and of remaining ignorant. More specifically, the aim is to show that weltanschauungen determine the initial fate of contentful ideas (their later fate is, as Popper holds, determined by observational test), to indicate the peculiar inaccessibility of weltanschauungen which helps to render them potent influences, and to consider the possibility of bringing them at least in some degree into the open. First to look at some elegant examples.

Some classic weltanschauungen: blinkers or spurs?

So long as the dignity of man required him and his habitat to be at the physical centre of the universe, it would be difficult to conceive of, to take seriously, and not to be automatically opposed to a weltanschauung that saw the Earth as merely one planet among others.

For all his immense successes against Cartesian physics, Newton lost to them over one encounter, which concerned the velocity of refracted light on entering a denser medium. Here it would seem that Newton's partiality for a particle weltanschauung prevented him from appreciating the strength of the Cartesian position. (Descartes found the sine law though he got the velocity wrong; it was got right however by Huyghens.) The controversy is complicated — a blend of observation, theory, and weltanschauung, which has been disentangled by Sabra (1967). Here I will mention only one fascinating detail. Newton, while able to derive the Snell-Descartes sine law of refraction, was committed to an increase of velocity in the denser medium. As Sabra (1967, pp.296, 314) makes clear, this was a consequence of the corpuscular weltanschauung, which stemmed from his atomism.

It is worth noting that Newton, who accepted the procedure of explanatory hypotheses with empirical content, repudiated hypotheses of a deeper kind, which he regarded as beyond the reach of observation and as irrelevant; he regarded them as mere imaginative productions. Sabra (1967, p.342) amusingly comments: 'Fortunately his fertile imagination was not hampered by his low opinion of imaginative products.' The story of refraction contains a wealth of example, but brief illustrations are easier to extract from Agassi's work.

Without a weltanschauung permitting of the hiving off of molecules from atoms, it was impossible for chemists early in the nineteenth century to grasp the chemical difference between salt-water and sodium chloride — between mixture and compound — hence they ran into difficulties about the universality of the law of definite proportions (Agassi, 1963, p.588).

There has been a fair amount of discussion, disentangled by Agassi (1963, pp.67—74) of the extraordinary scene when Oersted discovered electromagnetism, just after the end of a lecture-demonstration to a distinguished audience but before anyone had left the hall. Let us list the main points:

Oersted was convinced there was a connection between a current in a wire and a magnet placed beside it.

He quite correctly, in accordance with Newtonian theory, placed the wire in the east-west direction; there was no effect.

He quite reasonably thought this failure was due to too weak a battery; and he tried out stronger and stronger batteries.

He held that there are only a few kinds of matter but many types of manifestation of force: fairly large electric forces transform into heat; larger ones in thin wires transform into light. Agassi's meta-conjecture: did Oersted conjecture that still larger electric forces would transform into magnetism?

According to Agassi's reconstruction Oersted had three choices: (a) the long weak magnet (that is, the wire with current but poles unidentified) was too weak, (b) was not a magnet at all, or (c) the Newtonian theory yielding the east-west direction was false.

The moment after the conclusion of his public lecture Oersted is recorded as suddenly placing the wire in the north–south direction — and the famous electromagnetic phenomenon was found.

There are two questions: why did Oersted fail for so long to get the right answer; and why did he suddenly, without some appreciable length of reconsideration, hit on it?

It is essential to realise that for many *months* or even *years* Oersted had been trying to bring about the phenomenon and never once tried any direction other than east-west. This is not the behaviour of someone doodling, nor even of a Baconian inductive experimentalist; it is the behaviour of a man in the strong grip of a theoretical approach. It seems evident that he made no mistake over the various empirical hypotheses involved. Where he went wrong was in living within the framework of approach of Newtonian theory — that is, the weltanschauung. I would take it that the empirical Newtonian theory was converted into a weltanschauung. (Agassi calls him a Kantian and an apriorist.) How great was that influence!

This is the relevant point in the present context, but we may reflect a moment on the second question, which Agassi's reconstruction answers. To try some other direction for the wire in a fit of desperation would be thinkable for a small boy who looks on laboratory work as doing haphazard experiments, as trying out anything randomly, his only hypothesis being that *something may happen*; and it is thinkable for a Baconian inductive experimentalist (it would be interesting to know if one has ever existed or if anyone would offer such a worker a job), but apparently not for Oersted. He was obviously feeling desperate. Agassi rightly says he could hardly in a flash have questioned the Newtonian weltanschauung. So Agassi thinks the only alternative open to him was to wonder whether Oersted had made a mistake (which of course he had not) in his deduction from the Newtonian theory of the east-west direction. I think Agassi is right. At such a moment, there would be no point in trying some small deviation from that direction, so he would try the really contrasting possibility. He would accuse himself of a mistake (a perfectly correct thing to do) before questioning the weltanschauung (which most people, improperly, cannot consider until over-forced to do so). It takes ten times as much incompetence as would reasonably justify dismissal before a civil servant or a university teacher can actually be got rid of for incompetence, and likewise ten times as much reinforcement of hard knocks before fine intellectuals can admit weakness in their weltanschauung; and this is what conservatism and establishment consist of. The order of questioning is right (and of course the smart student, questioning the weltanschauung before he has learnt the value of the operations conducted within it, is wrong), but inflexibility, placing the weltanschauung inside the shrine until it is blown open, is wrong or at least slows up new thinking tenfold. Given that a weltanschauung must be walled off within a simulacrum, an innovator can break new ground only if immoderately persistent.

It is proper to add that Agassi could not have explained Oersted's failure and success without replacing the then existing weltanschauung of the history of science by his own new one.

The next example is Faraday's concept of a field as developed from an idea of Wollaston's. As Agassi (1971, pp.91f) depicts it, iron filings round a conductor are magnetised; if the iron filings are removed we suppose there remain a host of invisible magnets, and we may conceive of these as reduced in size towards zero so that their north and south poles practically coincide, leaving us with a near immaterial dipole (for the material has shrunk towards zero); reduce their size to zero and increase their numbers; and this gives us a Faradaic magnetic field. When reduced to zero the dipoles have no polarisation left; polarisation is represented, not by bands of iron filings, but by geometrical lines. This is foreign to the particle-minded, atomistic Newtonian ('it took physics three or four generations to cope with the upheaval caused by Faraday's collapse of a magnetic dipole into a point of polarization axis' (Agassi, 1971, p.93)). It replaces magnetic matter by a 'field'. A 'field' thus conceived is matterless.

This concept involves the concept of *forces in empty space*.

Now the Newtonian weltanschauung involves, notes Agassi (1971, p.109), only space, time, matter, and its properties; *and forces are properties of matter* — and could not conceivably within this weltanschauung be properties of space. (And on top of this, he required a new principle of conservation of force.)

Even Maxwell, steadfast supporter of Faraday, had difficulty in conceiving of force moving without a medium; but it was Maxwell's own attempt and failure to combine the movement of force with mechanism that led him and others to abandon mechanism. Hertz, too, for all his attachment to mechanism, concluded in the end that 'electric forces are polarizations *existing independently in space* . . . [they] are not simply parts or attributes of their causes . . . [we denote] them as polarizations. Whatever the nature of these polarizations may be' (see Agassi, 1971, p.163). Agassi suggests that Faraday had found an entity for which there was no existing category. He is concerned to highlight the traditional philosophical view that force must be bound to matter and thus to show the influence of that tradition on the history of physics. According to traditional philosophy, matter was an explanatory concept; with Faraday it became something to be explained — and he explained it, as it were, as an occlusion (or condenser) in a field of force (anticipating Einstein).

The use of reins

Earlier I have put forward the view that a weltanschauung prescribes what shall count as science and proscribes what shall not. The examples just given amplify: certain doctrines are not accepted by colleagues, some disregarded because they do not conform to 'science', that is, as we now see to a certain weltanschauung; more, an innovator has to struggle not only against his colleagues, he has to struggle against the weltanschauung he has been nurtured on — no easy task, since he does not know what it is until he has found an alternative. It would be a mistake to lay too much stress on the inhibiting effect of a weltanschauung — though we see only too much of this; we ought also to see the liberating effect. Maxwell is a good example of one liberated by Faraday's weltanschauungen.

If a weltanschauung can be used to blinker a scientist or to spur him on by giving him new policies, the blending of these two effects is to be seen in the use of reins. For reins can be used to reinforce blinkers or to encourage a wild dash — or even to blaze a new trail — exercising subjugation, encouragement, or controlled enterprise. But they cannot be dispensed with, as the hard-nosed scientist would wish and even believes at times has been done.

It is common enough for readers of history to become exasperated with the 'blindness', meaning the 'stupidity', of those who would not see. Inspection of their illustrious names shows they were not just a run-of-the-mill type. True they were usually no more than distinguished or outstanding — but they were at any rate of that calibre. If it takes, as it usually does in the initial stages, a man of the same superstar calibre as the unappreciated innovator — as it took Maxwell to appreciate Faraday — the matter is hardly one of mere stupidity. Besides, in a few generations, sometimes even in one, the young can take the old innovator in their stride. So it is a question of framework of approach within which one sees the issue. In short a weltanschauung can function as blinkers obstructing the vision even of those with high visual acuity.

A weltanschauung as a threat

A person can tolerate losing a new pair of gloves, having his new car destroyed, his old master stolen, perhaps the loss of his job, suffer extremely over the loss of a close member of his family, be undermined by revolution in his country, feel in exile that he has lost part of himself; but what if one of his weltanschauungen, say the belief in atomistic mechanism, forming a significant part of his total weltanschauung, is shattered? Part of his weltanschauung may be damaged, perhaps irreparably, when his country is in the throes of revolution or if he lives in exile; but it is about scientific weltanschauungen I am writing here. In *Too True to be Good* Bernard Shaw depicts the despair of an atheist who has lost his faith in atheism. Somewhat smart in execution perhaps, but the phenomenon shows very aptly what high-level scientists are up against when a Faraday comes along. It is not stupidity; it is the shattering of part of their world (with conceivably the risk of splitting their personality). With a novelist's skill one might bring this out graphically and fairly fully in a few hundred pages; here, lacking this skill, I will sum it up as 'no longer knowing where one is'.

When a man modifies his weltanschauung, it is usually a slow process, so that he adapts to the change gradually. There may be unending conflict, and usually is, for he usually retains unresolved a part of the discarded weltanschauung. So there is clearly no point in demanding even of highly intellectual scientists that they should engage in rational discussion; if they are equal to exploring they can discuss, but it cannot be forced.

I doubt if Faraday was intellectually cleverer than *all* his distinguished contemporaries. He may have been, perhaps a little. Or he may not. I suspect it is not the point. Once a scientist has the high level of intellect required for his work (there are many who forget this basic necessity), I suspect that it is capacity to entertain, on his own, alternative weltanschauungen that puts him in the superstar category (provided he does not break down).

Living in such a world, or worlds, puts such a scientist in the position of being alone. This is partly a natural practical consequence of the fact that colleagues cannot face going with him. If they can, that will mitigate his sense of isolation. But even so, he may still be burdened by the possibility of disaster — what if his dawning weltanschauung lacks all substance? He has to face this open outcome, with no guarantees, whether or not some colleagues join him.

It is even that he has no guarantee that he is not mad. One who looked madness in the face was the first to entertain the idea that the Earth is not flat (the flat-earth belief was surely only good commonsense). Another who must have looked madness in the face — saved by his confidence plus his modesty and his inherent capacity to accept the situation if he was wrong — was Einstein. How could any sane man suggest that the relative velocity of light, no matter what speed the other object was travelling at, was constant! (No one has commented on the surprising fact that his 1905 paper promulgating the theory of special relativity was accepted for publication — would it be accepted today?)

It has been pointed out to me by Synge (J.L., personal communication) that philosophy of science has thrown no light on the excitement of scientific work; I was impressed by the challenge, and think I can make a small contribution about it at this point. While not underestimating the thrill of finding a new generalisation or finding out how something works, like tracking down the way bees use visual cues to find their way about, I wonder whether the committed scientist is not rather trying to uncover nature's secrets, to find out at the deepest level what the world is made of and how it works,[1] presupposing that he comes to grips with philosophical foundations underlying scientific enquiry. Hence the fascination of weltanschauungen. By this is not meant philosophy according to philosophical schools, but philosophy as part of the life-blood of science itself, yielding what I have called the embedded ontology of science. Science in the sense of content-enquiries took over when philosophy seemed bankrupt; but this was because philosophy took another fork, so that its aprioristic cosmological ontology severed all connection with the ontology of science (not that there was anything wrong about that pursuit, only in its exclusiveness); but this attempted self-sufficiency of philosophy — inevitably evoking self-sufficiency on the part of science — should not blind us to the fact that science is pervaded by its own philosophy. All I have attempted here is to give a specific form to the philosophy underlying science, the weltanschauung (and the embedded ontology this attributes to the world), which makes it a trifle more understandable why this should stir the scientist who wants to grasp the world.

Relative weights of content and weltanschauung in enquiry

It is routine to check up on content hypotheses; for this the Popperian metascience of testing (falsifiability) by observation provides what is required. It is a disappointment for the investigator when a nice hypothesis fails. No worse. It has been alleged (there is some reason to think the story untrue) that Galileo first got the law of falling bodies wrong, thinking the velocity proportional to the height, but found out his mistake and realised that the velocity was proportional to the time. If he had made this mistake, and even if he had never corrected it, he would have made, in the idiom of the scientific jargon of today, a breakthrough, far exceeding in importance the matter of getting the details right or wrong. Many content hypotheses are of great practical importance and some — a few — of theoretical importance; so progress can be made by adding to them. They are relatively easy to challenge, and a well constructed challenge can be dealt with, specifically investigated, and the outcome decided.

It is otherwise with weltanschauungen.

They are hard to recognise.

They are hard to formulate.

There is no logical or observational means of testing them.

They may foster discovery, but can never be corroborated.

Though they may on occasion, rather fortuituously, be refuted (refuted by a satisfactory new theory), we discover their falsity without having a corroborated replacement, only a viable replacement.

You can specify how you would test, try to refute the content of a scientific theory; you cannot specify how you would attempt to challenge a weltanschauung.

It would be true to say that a weltanschauung is challengeable only when it has been challenged. However that alone is at least an entry point: we know that a weltanschauung is challengeable — difficult, fortuitous, the chance development of a refuting theory, but challengeable.

What this slender fact does for us is to encourage us to *seek out* our weltanschauungen even if we can do this only very partially, to *formulate* them as well as we can (without trying for too much precision), and to *entertain vague ideas* that might oppose them. If the reader thinks he is being asked to think about something neither he nor anyone else has ever thought about, let him not be put off. Something like this constantly takes place with every modest discovery the humblest makes. Breaking through to a new weltanschauung is like jumping in at the deep end unable to swim, or like learning to ride a bicycle. No one can tell you how to coordinate your movements so as to succeed in riding; you just try. You fall off, try again, and after a while you grasp the gestalt framework of balancing a little bit. This improves the framework, thence the balancing, and so on. Most people, confronted with a new task or a new problem in effect ask for directions how to proceed, not realising that the answer is a new creative endeavour — the only way to do something new is to try to do it. Tackling a new weltanschauung is in itself a creative endeavour.

Very often in the history of science the order of development appears to have been initially a new fundamental theory with empirical content to be followed by the dawning realisation of a new weltanschauung. Possibly Newton's theory of gravitation is such a one, and Darwin's theory of evolution, and Dalton's theory of the atom. Perhaps less often an awareness, possibly only dim, of the weltanschauung comes first and suggests new directions in which to seek new content theories. Probably Copernicus' heliocentric theory and Faraday's field theory fall under this heading. It may be, however, that the two are never really isolated in the scientist's mind; it may be that a glimmering of a new weltanschauung is present in approaching a content problem, and that the mists around the glimmering vision dissipate a little, and so on.

Whether or not the two are dimly present together in the scientist's mind, a likely process is that he makes a small dent in the current weltanschauung, hardly knowing what is taking its place. Fuller realisation takes time, and may not be articulated at all, leads to a larger dent and a further replacement, and so on. Years or a life-time it may take to reach a new weltanschauung in the round, sorting out where one stands vis-à-vis the old and the new and the conflict between them. (I do not know whether there is something comparable in mathematical development; it is characteristic of the great philosophical movements, even though there have been some outstanding philosophers who underwent no growth as regards their weltanschauungen.)

If encouraging the creative endeavour of trying to discern weltanschauungen means putting up with some cranks, then let us put up with them — better that than the boredom of the worst kind of routine 'research'.[2] Better a scientist muse in a hammock than conduct a senseless kind of research from 9.00 to 5.00.

In short, I am suggesting that in doing theoretical science we are not only seeking (functional) explanations, as Popper rightly holds, but also, following Hattiangadi, seeking descriptions of the unknown in the world, which I see in terms of improving our weltanschauungen to provide an improved scientific ontology. And the focal point for challenge in science should become our weltanschauungen.

Notes

1 Since writing this I have come across a comment on Galileo, by the historian Copleston, that he thought of mathematics 'as opening to us the very heart of Nature'.

2 The reader might indeed think I was taking the same line as Feyerabend here — accepting hobgoblins as well as scientists, even though I have criticised Feyerabend for this. But I think the discerning reader will see that there is a fundamental difference. Feyerabend is prepared to take theories of witch-doctors seriously — or at least that is what comes over. I am not. I am prepared — and have done so — to *listen* carefully for hours to cranks for the slightest hint of an 'interesting' idea (with the possibility of being put to work — I do not mind how 'mad' the origin of an idea if we can give it testable shape). (Putting up

with cranks means listening to lengthy diatribes which they lack the scientific education to blue-pencil.)

References

Agassi, Joseph (1963), *Towards an Historiography of Science*, published by *History & Theory*, Beiheft 2, Mouton, den Hag.
Agassi, Joseph (1971), *Faraday as a Natural Philosopher*, University of Chicago Press, Chicago.
Sabra, A.I. (1967), 'Theories of Light from Descartes to Newton', Oldbourne, London, pp.296, 314.

17 The handmaid – science or metaphysics?

It is of some significance that the distinctions between empirical content of a theory, embedded ontology, and weltanschauung are independent, if not of all, at least of certain basic controversial issues in the philosophy of science. There need be no significant disagreement among different schools of philosophy of science about the meaning of empirical content, however much they may disagree about criteria for discriminating it. I find it most convenient to adopt the criterion of testability by observation, but this is not essential. The intuitive distinctions between content, ontology, and weltanschauung would not be obliterated if some other criterion were used and could almost certainly be retained in some other form. Now the distinctions would seem to be neutral as between the notion of the objectivity of science on the one hand and that of incommensurability and relativity of truth on the other; neutral as between the doctrine of realism and that of conventionalism; as between the claim that 'truth' counts and that it does not; and as between the view of observation as theory-laden and the philosophy of observationalism.

Although many, or nearly all, of the arguments in the published controversies take place on tactical battlegrounds over logical, empirical, and theoretical points, the existence of weltanschauungen behind these is not hard to see once one is alive to the possibility.

There has arisen a wave of sentiment in the latter part of the twentieth century against objectivity. The sociology and psychology of it are subjects to be taken up elsewhere. But we may note such factors as disillusionment, the chill attributed to detachment, the lure of the personal, the need for self-justification. Again, 'truth' and 'reality' become irrelevant — irrelevant to the personal, how the person sees things, sees himself: *esse* is *sentire* and *esse* is *sentiri*.

Conventionalism means that you do not have to bother about reality as it really is, and therefore goes with denial of objectivity. Thus conventionalism and denial of objectivity fit the anti-rationalist weltanschauung just referred to. But to find that they are a weltanschauung is to remove any underpinning they may appear to have, and therefore to undermine their apparent self-evidence and persuasiveness and put the controversy in a different perspective.

As regards the philosophy of observationalism, it has long been seen, however dimly, to be a philosophy, but this recognition failed to weaken its influence — because of its appearing to be self-evident. Again to realise that it — and also its new rival, the theory-ladenness of observations — is a weltanschauung, is to introduce sobriety into the discussion of the issue of observationalism.

We are left with a certain choice, allowed by whatever doctrines are compatible with one another. 'Truth' and 'reality' cannot live in the same home with conventionalism. Conventionalism is incompatible with theory-laden observations (unless a way of overcoming this can be invented). So 'truth' and 'reality' have to live with theory-laden observations. But even to work out this position requires an excursion into the metaphysical question of whether or not there can be universal frameworks (for example, 'change occurs'). Rationality (versus incommensurability) involves at least such components as a universal framework, 'restricted realism', some theory of 'restricted reference', theory-laden observations, and of course relative falsifiability.

The polarisation in natural science is between, on the one hand, the possibility of relative falsifiability, a 'uni'-verse, commensurability, some objective reality, some possibility of reference, some 'truth', theory-domination based on tests, subservient observations — or, in a summary concept, rationality; and on the other hand the repudiation of all falsifiability including repudiation of the relevance of inconsistency, the assertion of a 'multi'-verse (not, that is to say, many in some relation to one another, but many that are unconnected or incommensurable), disregard of objective reality, of reference as illusory, and of 'truth' as an irrelevance, and instead domination by non-competitive ideas, that is, without discriminatory tests, by observations as meaningful only within a particular setting — or, in a summary concept, the horse without reins.

We can see that the question of which is the more basic, science or metaphysics (in the sense of the weltanschauung just indicated) ceases to be substantial: science cannot operate without reins; but not only does science presuppose the 'reins philosophy', summarily put as rationality, but it leads to that philosophy insofar as in doing science we seek knowledge of the world consisting of just those reins.

Never has the rationality of science, the philosophy of scientific reins been subject to such onslaught as at present. Of the components of rationality mentioned, each has been questioned not only by sceptical philosophers but some of them by very great natural scientists. I have been at pains to bring out that the undermining of rationality in the form of conventionalism has roots in Poincaré – which produced tremors in the first few years of this century but which is still with us. It is now even more disruptive since Kuhn (1962) gave new life to conventionalism. And this he brought about mainly in two ways: by undermining the role of falsifiability and by a sociology (presumably scientific)[1] of science. This movement assumed the form in Feyerabend, though not (or only reluctantly) in Kuhn, of the philosophy of the horse without reins.

It is perhaps odd that the total undermining of science, however much based on a misunderstanding of the role of relative falsifiability, should have come from a natural scientist. Kuhn's position has aroused the interest of most philosophers of science, though apparently not specially of natural scientists. Where his position really has told – and possibly done untold damage – is in the field of the social sciences. But not for the first time social scientists have drawn inspiration from a peripheral feature of natural science perhaps misinterpreting it as those following the rationality weltanschauung would hold; though more probably they have accepted Kuhn's thesis as valid because it comes from a natural scientist and therefore still commands the acceptance accorded to prestige; and perhaps they have adopted Kuhn's bare thesis as filtered through the intellectual community without noting Kuhn's own degree of caution. If natural scientists have taken only a passing interest in Kuhn's incommensurability form of conventionalism, several varieties of social scientists have found in it their salvation, and have found it as charismatic as a moth does the flame. Why?

Kuhn rightly noticed that scientists and philosophers of science had taken no account of the sociological factors involved in the development of natural science. In bringing these to light Kuhn at the same time undermines objectivity and truth. Their pedestal is occupied, according to Kuhn, by a new élite – the younger establishment of foremost scientists. But this is not the occupant of the pedestal that interests social scientists committed to people. The occupant for them is to be understood not in relation to rational standards but in the sense of an intellectual posit independent of standards – at least standards other than that of a humanistic, interpersonalist approach to people. Objectivity and truth are regarded as inimical to seeing things from the standpoint of the individual; the standpoint of the individual is outside the reins of objectivity and truth. Kuhn's corpus of scientists is, at any rate, an informed oligarchy. This sociology may be inadequate as philosophy of science, untrue to the realities of scientific development and to its logic; but even so, ironically enough its élitism would approximate for most of the time to those realities and that logic (though it might well be highly damaging when fundamentals are at stake). The social scientist's replacement of the scientific élite, which for Kuhn is a replacement of Popper's non-élite criterion of testability, is untrue to Kuhn and leaves him far behind in the interpersonalist revolt.

What seems to have been meted out to Kuhn by the doctrine of individuals as incommensurable persons is what happens typically in revolutionary movements: the charismatic inspiration — whether witting or unwitting in its author — is spread like a forest fire, while the author if he is not prepared to be more extreme than himself is pushed aside.

I think that the current fierce attack on science, objectivity, truth, and even rationality and logic, which may well be the fiercest ever mounted in history, is a good thing. This is because the theoretical tools, though rudimentary, are around for understanding the attack, for understanding genuine weaknesses in the position attacked, and for understanding a possibly healthy or valid aspect of the attack.

It is indeed true that natural science, from the time of its great revolution in 1543 to the present, has worked with a weltanschauung of what I have called *inert mechanism*, which in the social sciences is inappropriate, nearly though not quite useless, and enormously restrictive; but I have also indicated that that weltanschauung is not essential to science, though 'natural' enough to natural science, and that it can with no difficulty be replaced by a socio-personal weltanschauung enabling social science to be conducted as science (Wisdom, 1971).

It is also true that some social sciences have long respected the individual. Although he did not put it on a banner, Malinowski by his innovative fieldwork in anthropology took as his springboard the attitudes of his subjects as seen by themselves. His view of the individual goes back to almost the beginning of this century. And dynamic psychiatry stemming from Freud has done the same thing for nearly 100 years. Existentialism has not invented all this. But the movement of existentialism after World War II and also the post-phenomenological social-science movements have emphasised it. Perhaps they have been the first to realise that the classical weltanschauung of natural science has to be replaced — without of course realising that it has to be and can be replaced *within* the corpus of science — by a different weltanschauung for social science. And perhaps they have seen more vividly, even than is portrayed by Popper's thesis that observations are theory-laden, the depth of influence upon social observation by official and also by unarticulated social theories.

Thus there is distinct justification for the interpersonalist revolt, even after allowing for and correcting certain of its erroneous contentions.

On the other hand, the interpersonalist revolt, rightly annexing the general right to criticise, often fails to make any effort to understand science, or to accord to it the understanding on its own ground parallel to the way it demands that human beings must be understood by natural scientists from its interpersonalist point of view.

The conflict is plainly between two weltanschauungen, in capsule form the rationalist versus the philosophy of the horse without reins. Interpersonalist social scientists are keenly aware of this, though the notion and influence of weltanschauung has not been particularly noticed by philosophers of natural science. But interpersonalists, while using a new weltanschauung to attack an old one, show no sign of recognising that weltanschauungen are not provable and (by and large) not disprovable. So their confidence in the new one lacks an anchor. Yet, utilising one of the debating techniques of the last century or so, they may attempt to undermine science and the logic of consistency, because the weltanschauung of that logic can after all be replaced by the weltanschauung of the logic of dialectic. This position in social science, at least in part, comes from a misunderstanding of the Hegelian dialectic; for it is thought that the dialectic is an alternative to (and excludes) the ordinary logic of consistency. But Hegel never meant to assert that 'It is raining' is logically equivalent to 'It is not raining' — and you need not bother about your umbrella. And dialectic itself presupposes that in a dialogue you change your position as a result of opposition of some sort, that is, something is relinquished — or in short is accepted as false. One would appreciate it if an application of a principle, such as that of dialectic, were correctly (?logically) made.

Is there a criterion of weltanschauung for social science?

But we are still not at the end of the road. If two weltanschauungen clash — say self-realisation versus the horse without reins — and if there is no process of proof or disproof, are we totally without means of assessment? There are at least two possibilities relevant. There is the criterion of compossibility — to use Leibniz's term. Naturally there is the special case of logical consistency; but fully overt logical inconsistency that cannot be evaded with a modicum of adroitness is so rare that it is hardly worth considering. However, compossibility is usable. We can ask what a weltanschauung rules out, that is, what we have to give up to maintain it. On considering such expendable candidates, although we may not be able to disprove them, we may find them impossible to work without, and that would mean relinquishing the weltanschauung. It is usually a long haul to pursue this criterion in practice — evasion is too easy. Stronger is what I would call the enabling criterion: we ask whether a weltanschauung can do its job or gets in the way of its own goal. Perhaps this can be made use of here, bearing in mind the neutrality towards rival approaches of the distinction between empirical content and weltanschauung.

What I think has happened is this. Most revolutions in science are revolutions both of empirical content and of weltanschauung: sometimes the latter may have stimulated the former, more often the former ushered in the latter, though possibly they have developed together more generally than would appear. It is hard to think of an example of a natural science weltanschauung that has attracted no empirical content; but we might find examples in the ferment when scientists were baffled over the theory of heat or the theory of electricity in earlier days. If there is such, it would be dead. 'Normal' science proceeds to develop and proliferate in content without *change* of weltanschauung. Further, as is so blatant with some of the academic social sciences, scientific/empirical content without weltanschauung at all (except commonsense) is blind. On the other hand, a weltanschauung without empirical content is empty. Accordingly, I am suggesting that the interpersonalist revolt against science is a weltanschauung but is only a weltanschauung, that is, has no empirical content. Otherwise put, interpersonalists in revolt, however finely inspired though intellectually sidetracked, are purveyors of an ethos without grass roots. And to develop the grass roots of social empirical content oddly enough presupposes coming to grips both with science and with metaphysics as weltanschauung.

Unless we straighten out all these matters in philosophy of science, it is questionable whether we can hope to make progress with the philosophy of social science — and even more important, progress with social science itself.

Note

1 Presumably Kuhn would have to agree that his sociology of science is conventionalist, and not 'true' or 'objective' — whether he agrees or disagrees he is disembowelled by an obvious dilemma.

References

Kuhn, T.S. (1962), *The Theory of Scientific Revolutions*, Chicago University Press, Chicago.

Wisdom, J.O. (1971), 'Science versus the Scientific Revolution', *Philosophy of the Social Sciences, 1*, pp.123–44.

Index

Adams, 51
ad hoc, 45
Adler, 42
Agassi, 21, 28, 40, 62, 66, 67, 75, 109, 112, 135, 136, 142, 147, 148, 149, 159
Ajdukiewicz, 87, 120
algorithm, 13, 40, 73, see also induction
Anaxagoras, II, 36
Antisthenes, 10
approximations, 10e, 116
Aristotle, 9, 12
axioms, 46

Bacon, 10, 11, 12, 13, 37, 42, 43, 47, 83, 125, 138, 146, 148
Bakan, 122
Bartley, 136, 142
Bath, 79
behaviourism, 26
Bellarmino, 18
Berkeley, 10, 18
blinkers, see reins
blunt numbers, 116
Boas, 78

Bondi, 39
Bowers, viii, 14, 47, 59, 142
Bridgman, 19, 20
Bromberger, 135
building blocks, see observationalism
Bunge, 135
Burtt, 135
Butterfield, 62, 67

Campbell, 96, 99, 100
Carnap, 13
course, 18
Cohen, 127, 132, 133, 135
coherence, 111, 112
change, 119, 159
classification, 116, 119
concepts, non-instantive, 13, 95, 96, 99, 100, 101, 102, 103, 109, 113
conditional refutation, 4, 5, 16, 17, 20f, 32f, 67, 72f, 107f, 184, 187
conventionalism, 4, 5, 16, 17, 20f, 32f, 67, 72f, 107f, 184, 187

Copernicus, 26, 60, 77, 154
Copleston, 104
core, 118, 119, 120
correspondence, 112, 115, 117, 118, 119, 120
cosmological furniture, 118, 119, 120
Coulomb, 64
Craik, 100
creativity, vii, 5
crisis, 52, 53, 59, 63
criticisability, 136, 137, 138

Dalton, 163
Darwin, 163
data, see observation
decidability, 134
definition, see essentialism
demarcation-problem, 40, 42, 47, 53, 125, 138
denotation, 115, 117, 119, 157
Descartes, 17, 26, 95, 132, 147
Dewey, 28
dialectic, 160
dipole, 149
Duhem, 21, 108

Eddington, 94
eigen-value, see quantum mechanics
Einstein, 24, 42, 54, 55, 94, 111, 120, 142, 150, 151
embedded ontology, 127, 129, 130, 131, 132, 134, 137, 139, 156
empirical content, 131, 135, 156, 161
epiphinomenalism, 78
essentialism, 3, 4, 6, Ch 2, 8, 9, 76, 77, 79, 115, 116, 133
Euclid, 14, 24, 27, 110, 111, 143
experimentalism, 26, 50, 132; see factualism
explanation, 45, 46, 47, 100, 109, 150

Feyerabend, 89–90

Galileo, 17, 18, 19, 24, 26, 111, 113, 120, 121, 127, 128, 132, 152
generalisation, 13, 25, 34, 37, 100, 101, 102, 103, 105
Grünbaum, 28, 33, 121

Hall, 78, 80
Hamilton, 29, 122
Harvey, 97
Hattiangadi, ix, 29, 60, 101, 103, 120, 122, 150
Haynes, 14
Hegel, 160
Hempel, 100
Hertz, 149
heuristic, 12, 13
Hume, 10, 14, 42, 47
Hutten, 135
hypothetico-deductive, 35, 39

incommensurability, 111, 119, 120f, 157, 158, 159
induction, 11, 12, 13, 17, 25, 27, 34–8, 40, 43, 67, 71–4, 76, 102; see algorithm
inference-license, 28
instrumentalism, 16–18, 26, 28, 32, 34–7, 39, 72–4, 76, 77, 107, 113, 132
interpersonalism, 157–61

Jarvie, ix, 112
Jones, 135

Kant, 42, 47, 83, 148
Kaplan, 92
Kepler, 13, 26, 93, 102, 120, 127
key, 61–2, 67, 109
Körner, 116, 135, 143
Kuhn, 4, 21, Ch 7, 82, 111, 134, 141, 158, 159, 161

Lakatos, 4, 12, 38, 53, 54, 59, 60, 111, 120, 134, 135, 136, 141, 142
Lande, 19, 143
Lappin, 122
Leibniz, 142, 160
le Roy, 21, 29
Leverrier, see Adams
Locke, 14
logical positivism, 18, 20, 27, 28, 43, 98, 103, 120

McCormack, 7
MacKinnon, 29
Malinowski, 159
Mantle, 122
Marxism, 42, 43, 44
mathematics, 18, 19, 26, 41, 50, 111, 113, 117, 132
Maxwell, 101, 149, 150
May, vii, 142
mechanism, 8, 78, 79, 80, 159
Medawar, 39
metaphysics, viii, 3, 5, 6, 19, 20, 22, 27, 30, 43, 81, 82, 99, 125, 127, 134, 135, 136, 137, 140, 157, 161
metascience, 23, 25, 29, 36, 38, 42, 45, 47, 49, 51, 52, 57, 59, 60, 63, 65, 66, 67, 71, 79, 82, 91, 92, 93, 96, 99, 101, 111, 121, 130, 132, 133, 141, 152
middle ages, 3, 8, 12, 76
Mill, 12, 13, 28, 37, 125, 138
model, 140–1
Moore, 10
Munevar, 6, 7
Musgrave, 60, 86, 90

Newton, 17, 26, 42, 54, 63, 64, 67, 78, 93, 94, 101, 119, 127, 128, 129, 130, 131, 132, 133, 139, 142, 147
normal science, Ch 7, 63f
Nowell-Smith, 112

observation, see observationalism
observation, theory-laden, 4, 17, 18, 26, 71, 72, 77, 81, 82, 83, 84, 85, 86, 87, 89, 94, 96, 97, 98, 113, 114, 115, 118, 119, 120, 122, 125, 126, 143, 150, 153, 157, 159
observationalism, vii, 3, 4, 5, 6, Ch 2, 20, 21, 22, 26, 35, 36, 37, 38, 39, 44, 45, 47, 50, 51, 60, 71, 73, 74, 76, 77, 79, 80, 81, 82, 83, 84, 85, 86, 87, 88, 89, 90, 92, 93, 94, 97, 98, 100, 103, 110, 113, 116, 117, 118, 122, 125, 126, 127, 156, 157, 159
observation-refutable, 127, 128, 129, 130, 131, 132, 139, 143, 146, 152
Ockham, 10
Oersted, 147, 148, 149
ontology, 27, 29, 107, 127, 129, 133, 134, 139, 142, 152
operationalism, 17, 19, 20, 26, 75, 98, 132

paradigm-exploitation, 42, 53, 54, 55, 134, 137
Parmenides, 122
Popper, vii, 3, 4, 5, 10, 12, 14, 16, 17, 18, 19, 21, 23, 27, 28, 35, 36, 40, 41, 42, 43, 44, 45, 46, 47, 49, 50, 51, 52, 53, 57, 58, 59, 60, 65, 66, 71, 81, 82, 83, 92, 95, 99, 102, 107, 108, 112, 118, 122, 125, 126, 131, 135, 136, 138, 139, 141, 142, 143, 146, 148, 154, 159
phenomenalism, 16, 98
philosophies of science, new, vii
philosophy of science, role of, 6
Plato, 9, 12
Poincaré, 21, 22, 23, 24, 33, 108, 109, 110, 111, 122

pragmatism, 19, 27, 28, 98, 134
prescribing, 132, 133, 136, 137, 140
Price, 97
progress, scientific, viii
proscribing, 133, 136, 137, 140
puzzle-solving, 52, 63, 65

quantum mechanics, 17, 18, 27, 28, 128, 138, 143
Quine, 21, 113

Radknitzky, 6, 7
Ramsey, 100
reality, vii, 5, 18, 20, 21, 24, 25, 27, 32, 81, 111, 112, 113, 114, 117, 122, 138, 153, 157
refutability, *see* falsifiability
reins, 150f, 157, 158
relativism, 112, 118, 142, 156
Renaissance, *see* middle ages
research programmes, *see* Lakatos
restricted reference, *see* denotation
revolution in permanence, 65
Russell, 10, 97
Ryle, 28

Sabra, 143, 147
secularism, 79
Shaw, 151
Sheiham, viii
Skate, viii
Slate, 29
Skolimowski, 74, 83, 120
sluicegates, 63, 67
Snell, 147
sociological factors, 55, 57, 58
solipsism, 86, 87
spurs, *see* reins
structural, *see* induction
Synge, 73, 122, 151

Tarski, 38, 112, 113, 114, 116, 118, 119, 122
testability, *see* falsifiability
theory-refutability, 130, 133, 136, 139, 140, 143
Thompson, 135
Törnebohm, ix
truth, vii, 4, 11, 18, 19, 23, 24, 26, 43, 81, 103, 108, 109, 111, 112, 113, 114, 118, 126, 138, 156, 157, 158, 159, 160
truth-value, 4, 20, 22, Ch 4, 28, 122

Uranus, *see* Adams

value-judgement, 130

Watkins, 133, 136
weltanschauung, 79, 133f, 161